Guidance on the Personal Monitoring Requirements for Personnel Working in Healthcare

Series in Physics and Engineering in Medicine and Biology

About the Series

Series in Physics and Engineering in Medicine and Biology will allow IPEM to enhance its mission to 'advance physics and engineering applied to medicine and biology for the public good.'

Focusing on key areas including, but not limited to:

- clinical engineering
- diagnostic radiology
- informatics and computing
- magnetic resonance imaging
- nuclear medicine
- physiological measurement
- radiation protection
- radiotherapy
- rehabilitation engineering
- ultrasound and non-ionising radiation

A number of IPEM-IOP titles are published as part of the EUTEMPE Network Series for Medical Physics Experts.

Guidance on the Personal Monitoring Requirements for Personnel Working in Healthcare

Colin John Martin
Honorary Senior Clinical Lecturer, Department of Clinical Physics and Bioengineering, University of Glasgow, UK

David Temperton
Formerly Head of RRPPS, University Hospitals Birmingham NHS Foundation NHS Trust, UK

Anthony Hughes
Consultant Physicist, Christie Medical Physics & Engineering, Nuclear Medicine Department, Royal Preston Hospital, UK

Thomas Jupp
Principal Physicist, Regional Radiation Protection Service, Royal Surrey County Hospital NHS Foundation Trust, UK

IOP Publishing, Bristol, UK

ISBN 978-0-7503-2199-0 (ebook)
ISBN 978-0-7503-2197-6 (print)
ISBN 978-0-7503-2198-3 (mobi)

DOI 10.1088/2053-2563/aaddb6

Version: 20181101

IOP Expanding Physics
ISSN 2053-2563 (online)
ISSN 2054-7315 (print)

British Library Cataloguing-in-Publication Data: A catalogue record for this book is available from the British Library.

Published by IOP Publishing, wholly owned by The Institute of Physics, London

IOP Publishing, Temple Circus, Temple Way, Bristol, BS1 6HG, UK

US Office: IOP Publishing, Inc., 190 North Independence Mall West, Suite 601, Philadelphia, PA 19106, USA

Contents

Preface

Arrangements for personal monitoring have evolved as dose limits and practices using radiation have developed. The latest European and International Basic Safety Standards published in 2014, refined dosimetry requirements and contained a substantial reduction in the dose limit for the lens of the eye, as recommended by ICRP, and this has particular implications for practices in healthcare. Individual staff members working in hospital radiation departments use x-ray imaging, and some perform interventional procedures on patients, which require their presence close to the x-ray equipment. Staff also handle radioactive materials and deal with patients containing radioactive materials, which require them to work within a variety of radiation fields. Staff need to use both protective devices and personal protective equipment, so their exposure to radiation is generally not uniform. This presents particular issues for personal dosimetry. For example staff using x-rays wear protective aprons which shield their trunk, but their head and eyes are exposed to radiation, as they need to view the procedures being undertaken. Thus the new dose limit for the eye, which is now similar to that for the effective dose to the body, has become the limit that is more likely to be approached or even exceeded. In nuclear medicine, vials and syringes containing radionuclides which are handled by staff can give rise to large dose gradients across the hand, so that prediction of the dose to the fingertips is difficult. Therefore new approaches involving more personal dosimetry are required, and methods are needed that can be used to predict probable dose levels, so that risk assessments can be prepared to determine the level of dose monitoring for individual staff members. This book includes short reviews of dose levels reported in the literature to allow anticipated exposure levels within different clinical specialties to be predicted. It sets out recommendations, designed to help radiation protection practitioners and healthcare workers to assess exposure levels for personnel and determine monitoring requirements based on established rules. Specific recommendations for different staff groups are included in boxes for easy identification. Guidance on dose levels at which monitoring should be considered is included in tables. The implementation of robust systems for dose monitoring and arrangements for investigation when dose levels exceed pre-set levels based on recommendations given in this report should facilitate optimization of staff radiation protection.

Acknowledgements

The authors wish to acknowledge the contribution of Victoria Bassett-Smith and Jim Thurston to initial discussions of the composition of the document. Also helpful were comments on drafts of the document from Siobhan McVey, David Sutton, Andrew Hince, Philip Orr, Michael Watt, Debbie Peet and Tamar Willson.

Author biographies

Colin John Martin

Colin worked as a hospital-based Medical Physicist for 38 years. He joined the Medical Physics Department at Aberdeen Royal Hospitals in 1976 and was Head of the Health Physics Service for the West of Scotland, based in Glasgow, from 1995 to 2013. He is an honorary senior clinical lecturer for the University of Glasgow and has carried out research in many aspects of radiation protection and radiation dosimetry. He has been a certified RPA since 1990 and has been an RPA2000 Assessor since 2000. He is Vice-Chair of ICRP Committee 3, and a member of several ICRP Task groups, IAEA committees, and UK working parties. He has co-authored/edited several textbooks on radiation protection, written over 200 articles including 130 papers in peer reviewed scientific journals, and is a member of Editorial Boards for the *Journal of Radiological Protection* and *Radiation Protection Dosimetry*.

David Temperton

David was Head of RRPPS (University Hospitals Birmingham NHS Foundation Trust) from 1994 until his retirement in 2017. He was a consultant clinical scientist and Medical Physics Expert. He acted as Head of the RRPPS's approved dosimetry service (ADS) which was the largest such service in the NHS. He was a certificated Radiation Protection Adviser since 1993 giving advice to clients in the healthcare, industrial and academic sectors. He was an Honorary Lecturer at the University of Birmingham.

Anthony Hughes

Tony studied Medical Physics at the University of Aberdeen, obtaining a PhD in 1990. For the next 13 years he worked as a clinical scientist in radiotherapy at Clatterbridge Hospital and nuclear medicine at the Royal Liverpool Hospital. For the past 16 years he has been part of the regional physics service operated by Christie Medical Physics & Engineering, based at the Royal Preston Hospital. He has been an RPA since 2007 and an RWA since 2011, specialising in nuclear medicine and PET.

Thomas Jupp

Tom Jupp studied Medical Physics at the University of Surrey, achieving an MSc in 2008. He completed Clinical Scientist training while working in the Medical Physics departments in the Royal United Hospital Foundation Trust in Bath and in Gloucestershire Hospitals NHS Foundation Trust, being awarded the Association of Clinical Scientists Certificate of Attainment in 2011. He has since worked in the Radiation Protection Department at the Royal Surrey County Hospital NHS Foundation Trust as a Medical Physics Expert and obtained Radiation Protection Adviser certification in 2018.

IOP Publishing

Guidance on the Personal Monitoring Requirements for Personnel Working in Healthcare

Colin John Martin, David Temperton, Anthony Hughes and Thomas Jupp

Chapter 1

Requirements for monitoring radiation dose

1.1 Introduction—historical perspective

From the discovery of x-rays in 1895 and radioactivity in 1896 the use of such sources of ionising radiation in medicine rapidly became almost ubiquitous in many countries in Europe and in the USA by the early 1920s. During this period there developed an understanding that there could be deleterious effects from high exposure to this radiation. Skin burns were observed, and it was recognised that workers exposed to such high radiation doses were developing other acute effects, in some cases leading to death. However, at that time there was no means to measure the radiation exposure, and no system of units to quantify the dose received.

In the 1920s the International Commission on Radiological Units (ICRU) was established to develop ways to detect and measure ionising radiation, and to define a system of units for the quantities measured. The International Commission on Radiological Protection (ICRP) was also established to propose ways and means to restrict exposure to staff working with ionising radiation for their health and safety.

From the late 1920s through to the late 1950s, statutory limits on the radiation dose considered acceptable from occupational exposure were almost exclusively aimed at preventing short-term (acute) deterministic effects. Although since Hiroshima and Nagasaki there had been mounting evidence of longer-term stochastic effects—the risk of cancer or genetic effects—little consideration was given to setting limits on the basis of reducing the likelihood of such effects later in life.

The first dose limits intended solely for the purpose of reducing the stochastic risk of 'late effects', were proposed in 1958 by the publication of the first Recommendations of the International Commission on Radiological Protection (ICRP 1959). These were called Maximum Permissible Doses.

doi:10.1088/2053-2563/aaddb6ch1

In 1977, the ICRP updated those recommendations in Report 26 (ICRP 1977). This document set out to establish a more comprehensive framework for radiation protection, including the concepts of justification, optimisation and limitation, and the use of monitoring of radiation dose—both in the workplace and for individual workers. It lays the responsibility for the protection of workers and the public firmly on the shoulders of the radiation employer, in line with all health and safety regulation, and as such it is considered to be the start of the 'modern era' of radiation protection.

Three years later the European Commission published a directive on the basic safety standards to be applied to occupational and public exposure to ionising radiation (EC 1980), which in turn required every national government within the European Community to implement legislation to protect workers and the public from radiation exposure arising from such work practices. In the UK, these requirements were enacted under the Health and Safety at Work Act 1974 through the Ionising Radiations Regulations 1985 (IRR 1985).

The key points on monitoring in ICRP Report 26, the EC Directive, and the Regulations, were these:

- the workplace should be monitored;
- individuals likely to receive doses more than three tenths of the dose limit should be subject to systematic monitoring;
- other staff should be considered for monitoring depending on a risk assessment.

Since then the ICRP have updated the recommendations on two further occasions—firstly in ICRP Publication 60 (ICRP 1990) and secondly in ICRP Publication 103 (ICRP 2007). In both cases the main framework has been left unaltered, as has the emphasis on risk assessment in managing radiation hazards in the workplace, and establishing the need for monitoring the environment and workers.

1.2 Requirements for monitoring radiation in the workplace (1999–2018)

The adjustment of radiation risk factors in ICRP Report 60, together with other matters, led to a new EC Directive (EC 1996) which was subsequently enacted in the UK by the Ionising Radiations Regulations 1999 (IRR99). Following the latest ICRP Recommendations published in 2007, the EC has published updated Basic Safety Standards with a new EC Directive (EC 2013) which required EU States to enact revised legislation by 6 February 2018. Although the basic framework for radiation protection in the workplace has not changed significantly, there are a number of changes that have an impact on the requirements for monitoring individuals. In the UK these have been incorporated into the new Ionising Radiations Regulations 2017 (IRR17) which came into force on 1 January 2018.

Annual limits are specified for the effective dose, as defined by the ICRP (20 mSv), the equivalent dose to the lens of the eye (20 mSv), and the equivalent dose to the skin (500 mSv). The latter limit applies to the dose averaged over any area of 1 cm^2, regardless of the area exposed. There has been a substantial reduction in the dose limit for the lens of the eye following accumulation of evidence regarding the induction of cataract and revised recommendations from the ICRP (ICRP 2012). This has significant implications for personal monitoring in several areas of medicine.

Details of the legislative requirements in the UK as set out in IRR17 are given in this section.

Regulation 20 states that employers must ensure that designated areas (controlled and supervised areas) are monitored with the results recorded and kept under review. Records of such environmental monitoring should be kept for two years.

It also points out that the monitoring equipment used should be properly maintained and adequately tested at regular intervals under the supervision of a 'qualified person', who must be fully conversant with currently accepted standards of equipment testing.

Regulation 21 then states that employees who are likely to receive an effective dose of more than 6 mSv, an equivalent dose more than 15 mSv to the lens of the eye, or 150 mSv to the skin/extremity must be designated as 'classified' persons (also known as Category A exposed workers within the Directive). It is important to consider doses from both routine work and possible accidents when deciding whether an employee is likely to exceed these values. Employees must be at least 18 years old to be classified and must be certified fit for work by an appointed doctor or employment medical adviser. Classified persons can only be de-designated at the end of a calendar year unless the appointed doctor instructs to the contrary, or the employee is no longer employed by the same employer in a capacity which is likely to result in significant exposure to ionising radiation during the remainder of the calendar year.

The employer is required to make assessments of the dose to a classified person, as set out in Regulation 22. Arrangements should be made with an Approved Dosimetry Service (ADS) to make systematic assessments of doses and to maintain records. These dose records for classified workers must be available to the employer and kept until the wearer reaches, or would have reached, the age of 75 but in any event for at least 30 years from when they were made. A summary of the results of such monitoring must be provided to the employer at regular intervals. The requirements for approval of dosimetry services are set out by the Health and Safety Executive (HSE) and are summarised later in this chapter.

Employees who undertake work with ionising radiation but are not designated as classified do not need to be monitored individually. However, a system must be in place and be kept under review to ensure that they will not receive an exposure which may require them to be classified.

If an employee ceases to be classified, the ADS must send a copy of the 'termination record' (i.e. a summary of doses recorded whilst the person was classified) to the employer and the HSE. The ADS must also send to the HSE summaries of the doses received by all classified persons within three months of the end of each year. The Central Index of Dose Information (or CIDI) which receives these reports is now part of the HSE.

Employers must provide the ADS with suitable information about their classified persons, and where such classified persons are also 'Outside Workers' they must be provided with a radiation passbook (Regulation 22). Outside workers and radiation passbooks are discussed in more detail in chapter 9.

For all classified persons the employer must provide details of doses received, given reasonable notice. The employer must also maintain records of doses received for at least two years. If the dosemeter of a classified person is lost or damaged the employer must make an investigation and provide the ADS with a dose estimate for this period (Regulation 23, see chapter 9). If that is not possible the ADS must enter a notional dose which is the proportion of the annual dose limit for that period. If the employer has reason to believe that a dose entry is incorrect they must undertake an investigation and inform the ADS (see chapter 9).

In the event of an accident where an employee is likely to have received an effective dose exceeding 6 mSv, an equivalent dose more than 15 mSv to the eye or greater than 150 mSv to the skin/extremity dose, an investigation should be made (Regulation 24). The risk assessment should have identified that such an accident was possible, so the employee should already be classified and their dosemeters should be returned immediately for processing. If the employee was not wearing a dosemeter, the manager should consult with the RPA to arrive at a suitable assessment of the dose received. Records of accidents should be maintained for 30 years or until the individual reaches, or would have reached, the age of 75.

The HSE have published an associated Approved Code of Practice (ACOP) and set of non-statutory guidance designed to interpret the requirements of IRR17, including those for monitoring areas and workers, and to offer explanations of how these may be achieved (HSE 2018). The Institute of Physics and Engineering in Medicine also prepared Medical and Dental Guidance Notes (MDGN) as an interpretation of the requirements of IRR99 applied to the medical use of ionising radiations (IPEM 2002) and these have been updated to reflect IRR17 (Saunderson *et al* 2018). It should be noted however that these are non-statutory and as such have no legal status. The employer is encouraged, when considering the need for monitoring, to refer to both the ACOP and MDGN, in consultation with a suitable radiation protection adviser (RPA).

1.3 What dose quantity is measured?

Although the regulations define dose limits in terms of effective dose or equivalent dose (as defined by ICRP), personal monitoring is undertaken by assessing operational quantities which are easier to measure. These have been defined by ICRU (ICRU 1993). The quantity which is relevant to external personal monitoring is personal dose equivalent, $H_p(d)$, which is the dose equivalent in soft tissue below a specified point on the human body at an appropriate depth, d, below the position where the dosemeter is worn. A depth of 10 mm is usually employed for monitoring of effective dose or equivalent doses to other organs deep in the body and the personal dose equivalent is denoted by $H_p(10)$. Depths of 0.07 mm and 3 mm are employed (and denoted as $H_p(0.07)$ and $H_p(3)$) when monitoring doses to the skin and the lens of the eye respectively.

Dosemeters are worn on different parts of the human body and so the exposure to radiation scattered from adjacent tissue will vary. Therefore calibrations for $H_p(d)$ are carried out on appropriate phantoms made of tissue equivalent material that simulate the conditions in which they are worn in practice. For example a slab phantom (30 × 30 × 15 cm) is used for $H_p(10)$ to assess whole body doses, a right cylinder phantom (20 cm diameter and 20 cm height) is used for eye dosimetry in terms of $H_p(3)$, and a rod phantom (a cylinder of 19 mm diameter) is used for extremity finger doses in terms of $H_p(0.07)$. The operational quantity is obtained through calculation, utilising Monte Carlo simulation of radiation interactions within mathematical models of the relevant phantoms to convert air kerma to the relevant quantity. Since dosemeters are worn over or under protective lead aprons in radiology, this will change the exposure conditions, and so add additional uncertainty especially for assessments of superficial doses (Quintero-Quintero *et al* 2018).

For assessments of committed effective dose the operational quantity is the intake of radioactive material in Bq. In particular, the annual limit on intake (ALI) is the activity of a particular radionuclide which will give a committed effective dose (over 50 years) of 20 mSv.

1.4 Approval criteria for dosimetry services

Over and above any requirements set out within the regulations themselves, the HSE publish a set of criteria to be achieved by Dosimetry Services seeking Approved status in the UK. There are criteria for monitoring external radiations (HSE 2008), internal radiations, and for coordination and record keeping that will be updated to align with IRR17. These include aspects of reliability, consistency and security of the dose data held.

The dose assessment service should have a quality assurance programme in place to ensure that results are accurate, conform to criteria for approval and are retrievable. The dosemeter used must have a consistent level of performance in the radiation fields and ambient conditions likely to be encountered when being used in practice. The expected dose range is ~0.1 mSv to ~1 Sv for gamma and beta radiation and ~0.2 mSv to at least 50 mSv for neutrons. The calibration should be traceable to national standards and the service should be aware of the uncertainties associated with the calibration. Doses should be assessed in the operational quantities discussed in section 1.3 and reported to the record keeping service within 14 consecutive days. There must be arrangements in place for reporting late dosemeters. The service should be able to re-assess doses for up to two years generally (and five years for persons subject to a five year dose limit). For thermoluminescent dosimetry systems (see chapter 2) this requirement is achieved by storing the glow curve during readout. The service must undergo regular performance tests. In order to pass these tests the overall bias must be less than 20% and the relative standard deviation (RSD) must be less than 10%. There are also specific bias and RSD requirements for specific groups of dosemeters (HSE 2010).

Employers need to have an accurate indication of the total radiation exposure of their radiation workers. An individual radiation worker may be monitored for whole body irradiations from photons and neutrons and for internal radiation. All of these will contribute towards the effective dose. In addition, the wearer may have worn dosemeters to check doses to the extremities or eyes. These dose assessments may be carried out by a variety of approved dosimetry services. The service approved for coordination and record keeping ensures all measurements for classified persons are collated in secure and clear records containing a very comprehensive list of items specified by HSE. The services must:

- have a system which prevents corruption or degradation of data;
- ensure dose information is transferred into the records within 10 working days;
- exchange information with the Central Index of Dose Information (CIDI);
- inform the employer when an estimated dose is required because, e.g. a dosemeter has been lost or damaged;
- flag and notify the employer regarding doses above specified action levels;
- alter the dose record only in authorised circumstances;
- maintain the records securely until the wearer reaches the age of 75 but, in any event for 30 years from when they were made;
- provide the employer with regular clear written dose summaries and whenever an employee leaves their employment.

1.5 Other standards applicable to dosimetry services

There are a number of international standards that are applicable to work carried out by a dosimetry service. Indeed in some countries in Europe and further afield, these standards form the basis for 'approving' dosimetry services by the national authority/regulators for those countries. Details of some of those standards are given in chapter 2.

1.6 Studies of personal dosimetry for medical staff

As the application of personal dosimetry in healthcare has evolved, studies have been reported in the literature on an ad hoc basis. Approaches in different parts of the world have varied, but the basic requirements have followed recommendations made by the ICRP. A large-scale study was undertaken with funding from the European Atomic Energy Community between 2007 and 2011, involving nine European countries, called ORAMED: optimisation of radiation protection for medical staff (Vanhavere *et al* 2012). This collected data from dosimetric measurements was made on many different techniques and has provided valuable data from which guidelines can be developed. These guidelines have been developed taking account of results from the ORAMED and other studies. Brief descriptions are included of background studies on which specific recommendations are made. Results from the ORAMED report are referred to extensively throughout this document, but it has to be borne in mind that techniques employed in other countries are not necessarily those used and applicable in the UK, and recommendations made will be those considered to be the most appropriate for adoption in the UK.

1.7 Summary

The Recommendations of the ICRP have evolved over the past 40 years, leading to new Basic Safety Standards, which in turn have led to EC Directives requiring EU member states to enact legislation to bring the recommendations into law.

It is compulsory under the current regulations to assess the doses of individuals who have been designated as classified persons. The key reasons for carrying out individual personal monitoring on non-classified workers:

- firstly, monitoring workers who regularly receive measurable doses which are a significant proportion of the threshold which would require them to be designated as classified;
- secondly, monitoring a group of workers for which the results of a radiation risk assessment indicate that no-one is expected to receive a significant dose, but there is a need to verify and regularly review the results of that risk assessment;
- thirdly, monitoring a group of workers who are not normally expected to receive anything other than an insignificant dose but who work near to, or occasionally in, an area where, in an accident situation, they may receive a significant dose;
- finally, the employer may choose to monitor non-classified staff who are entering controlled areas in order to demonstrate that their doses comply with IRR17 Reg 19 or to reassure staff (including pregnant workers) that their doses are acceptably low.

In the following chapters of this report, these considerations for monitoring are applied to the range of routine uses of ionising radiation in a medical setting—from standard planar diagnostic x-ray imaging, through to CT scanning, interventional radiology, diagnostic and therapeutic nuclear medicine, to external beam radiotherapy and brachytherapy. One of the biggest challenges associated with personal monitoring of staff in the healthcare sector is ensuring that staff reliably and regularly wear their dosemeters correctly and return the dosemeters for assessment at the end of the issue period. Chapter 8 discusses the need for robust management systems controlling the use of personal monitors.

References

European Commission 1980 Council Directive 80/836/Euratom laying down basic safety standards for protection of the health protection of the general public and workers against the dangers arising from ionising radiation *Off. J. Eur. Commun.* L246 http://eur-lex.europa.eu/legal-content/EN/TXT/PDF/?uri=CELEX:31980L0836&from=EN (accessed May 2017)

European Commission 1996 Council Directive 96/29/Euratom laying down basic safety standards for protection of the health of workers and the general public against the dangers arising from ionising radiation Off *J. Eur. Commun.* **39** L159 http://eur-lex.europa.eu/legal-content/EN/TXT/PDF/?uri=CELEX:31996L0029&from=EN (assessed 26/4/16)

European Commission 2013 Council Directive 2013/59/Euratom of 5 December 2013 laying down basic safety standards for protection against the dangers arising from exposure to ionising radiation, and repealing Directives 89/618/Euratom, 90/641/Euratom, 96/29/Euratom, 97/43/Euratom and 2003/122/Euratom *Off. J. Eur. Union* **57** L13 http://eur-lex.europa.eu/LexUriServ/LexUriServ.do?uri=OJ:L:2014:013:0001:0073:EN:PDF (assessed 26/4/16)

HSE 2008 *Requirements for the Approval of Dosimetry Services under the Ionising Radiations Regulations 1999. Part 1—External Radiations* (Bootle: HSE) http://www.hse.gov.uk/radiation/ionising/dosimetry/requirements-pt1.pdf (accessed 27/2/2018)

HSE 2010 *HSE Statement on the Approval of Dosimetry Services* (Bootle: HSE) http://www.hse.gov.uk/radiation/ionising/dosimetry/dosimetry-state.pdf (accessed 27/2/2018)

HSE 2018 *Work with Ionising Radiation. Ionising Radiations Regulations 2017, Approved Code of Practice and Guidance* Publication L121 2nd edn (London, UK: Health and Safety Executive, The Stationery Office) http://www.hse.gov.uk/pubns/books/l121.htm (accessed 18/9/2018)

International Commission on Radiation Units and Measurements 1993 Quantities and units in radiation protection dosimetry *ICRU Report* 51 ICRU, Bethesda, Maryland

ICRP 1959 Recommendations of the ICRP. ICRP Publication 1

ICRP 1977 Recommendations of the ICRP. ICRP Publication 26 *Ann. ICRP* 1

ICRP 1990 Recommendations of the ICRP. ICRP Publication 60 *Ann. ICRP* **21** 1–3

ICRP 2007 The recommendations of the International Commission on Radiological Protection. ICRP Publication 103 *Ann. ICRP* **37** 2–4

ICRP 2012 ICRP Statement on tissue reactions and early and late effects of radiation in normal and health tissues—Threshold doses for tissue reactions in a radiation protection context. ICRP Publication 118 *Ann. ICRP* **41** 1–322

IPEM 2002 *Medical and Dental Guidance Notes; A Good Practice Guide on All Apsects of Ionising Radiation Protection in the Clinical Environment* (York: IPEM)

Quintero-Quintero A, Patiño-Camargo G, Soriano Á, Palma J D, Vilar-Palop J, Pujades M C, Llorca-Domaica N, Ballester F, Vijande J and Candela-Juan C 2018 Calibration of a thermoluminescent dosimeter worn over lead aprons in fluoroscopy guided procedures *J. Radiol. Protect.* **38** 549–64

Saunderson J R, Mayles W P M, Mairs W, Rowley L and Worrall M 2018 *Medical and Dental Guidance Notes. A Good Practice Guide on all Aspects of Ionising Radiation Protection in the Clinical Environment* 2nd edn (Bristol: IoP Publishing Limited)

The Ionising Radiations Regulations 1985 *SI 1985 No. 1333, London HMSO*

The Ionising Radiations Regulations 1999 *SI 1999 No. 3232, London HMSO* http://www.legislation.gov.uk/uksi/1999/3232/contents/made (accessed 9/2/18)

The Ionising Radiations Regulations 2017 *SI 2017 No 1075, London HMSO* http://www.legislation.gov.uk/uksi/2017/1075/contents/made (accessed 9/2/18)

Vanhavere F *et al* 2012 ORAMED: optimisation of radiation protection for medical staff *7th EURADOS Report* 2012-02 ISSN 2226-8057, ISBN 978-3-943701-01-2. Available at: http://www.eurados.org./en/Documents_Publications

IOP Publishing

Guidance on the Personal Monitoring Requirements for Personnel Working in Healthcare

Colin John Martin, David Temperton, Anthony Hughes and Thomas Jupp

Chapter 2

Dosemeters available

2.1 External monitoring

2.1.1 Introduction

Routine personal monitoring can involve active or passive systems. Passive dosemeters such as thermoluminescent (TL) or optically stimulated luminescent (OSL) dosemeters are worn for extended monitoring periods (e.g. one, two or three months) and the integrated dose received is read out at the end of this period. An active system gives an immediate reading of the dose, and possibly the dose rate, and is particularly suited to short monitoring periods and for high dose situations when it is combined with an alarm feature. Either type of dosemeter needs to be as easy to wear as possible with clips that attach to e.g. clothing, spectacles, lead aprons etc that ensure it is secure whilst being worn.

2.1.2 Whole body monitoring

2.1.2.1 Passive whole body dosemeters

2.1.2.1.1 Photon and β-radiations

Whole body dosemeters or badges usually consist of a holder with different filter areas and a robust card to which the various dosemeter elements (or detectors) are fixed. The performance of each system depends on which detector materials are used, the size (particularly thickness) and form of the detector, the filter holder and the readout process. The dosemeter card will have two or more detector elements.

- One detector is placed below a very thin filter so that it can measure $H_p(0.07)$.
- The second detector is below a thicker filter designed to enable it to measure $H_p(10)$.
- An additional detector below a different filter can provide information on the energy of the incident radiation. (Choosing a detector material with a more

doi:10.1088/2053-2563/aaddb6ch2

pronounced energy response can help distinguish between different radiation energies.) Algorithms can be used to compare the responses from the different elements to evaluate the energy and can then apply any necessary corrections for the energy responses of the detector material to provide a more accurate assessment of the dose.
- A pair of detectors to give neutron dose assessments.

When luminescent materials are exposed to a dose of radiation they store some of the energy from the exposure. This stored energy can be released after exposure by stimulating the dosemeter using an external source of energy. The amount of luminescence is proportional to the applied dose. In TL dosimetry, heat is applied to stimulate the phosphor. LiF:Mg,Ti or high sensitivity LiF:Mg,Cu,P are the most common TL materials used for personal monitoring but other materials such as $Li_2B_4O_7$:Cu and $CaSO_4$:Tm are also employed. The heating cycle of the dosemeter must be precisely controlled to achieve reliable and consistent dosimetry results. In OSL, crystalline aluminium oxide (Al_2O_3:C) or sintered ceramic beryllium oxide (BeO) are used as the luminescent material. The detector is stimulated with pulses of blue or green light using lasers or LEDs. The stimulation technique has been optimised by carefully choosing the pulse width, the pulse power, and the timing sequence. In OSL the readout process does not alter the crystal, so it can be reread many times to improve the accuracy of the result, unlike TL dosemeters (TLDs). In both cases, the energy is released in the form of light which can be measured, for example with a photomultiplier system, and calibrated in terms of the dose received. Some OSL dosemeters include a filter pack imaging area which can help determine conditions during exposures such as whether the dosemeter was moving during the exposure. This analysis is only feasible for energies below 150 keV and for reasonably high assessed doses (e.g. around 1 mSv).

In the UK, only TL and OSL dosimetry systems are used but in other parts of the world, other systems such as radiophotoluminescent glass are employed.

Performance figures are given by the manufacturers of the various systems and differ slightly. For photons, the energy range is typically from around 5–15 keV to 5–7 MeV for $H_p(10)$ whereas for $H_p(0.07)$ the lower energy limit is similar but the upper energy limit varies from 1.25–6 MeV. Some whole body dosemeters are also approved for assessing $H_p(3)$ to the lens of the eye from photons and have a similar energy range. For β-particles, the E_{max} energy range, when assessing $H_p(0.07)$, is more typically 700 keV to 2.2 MeV. The measured dose range is usually from at least 0.1 mSv (and can be as low as 0.01 mSv) up to 10 Sv.

2.1.2.1.2 Neutrons

Photons are always present in neutron fields but the ratio of neutron to photon dose equivalent can vary by orders of magnitude so the neutron component cannot be derived simply from a photon measurement. The radiation weighting factor for neutrons above 10 keV varies between 2.5 and 20 with the peak occurring at 1 MeV (ICRP 2007). For work involving neutron fields, a neutron dosemeter needs to be worn in addition to a photon dosemeter. In practice, personal monitoring for neutron fields usually needs to cover two energy intervals described as 'fast' and 'slow' neutrons. An exposure to the unshielded source of neutrons (the head of the LINAC or the target/beam tube of the cyclotron for example) will usually result in an exposure to fast (more energetic) neutrons. An exposure to a scattered neutron field in which the neutron energy has been reduced by scattering collisions in some shielding or moderating material would result in an exposure to slow neutrons (within the maze of a radiotherapy bunker for example).

Various different types of dosemeter are available for neutron dosimetry. TLD systems use TLD chips containing ^{6}Li and ^{7}Li. Although these two elements result in a similar photon response they give very different responses to neutrons. ^{6}Li gives a high response to neutrons, particularly very low energy thermal neutrons whilst ^{7}Li has a low response for neutrons of energies typical in shielded and/or highly scattering environments. It is therefore possible to use a combination of ^{6}Li and ^{7}Li TLD chips as a neutron personal dosemeter which also provides some neutron/photon discrimination. In the so-called albedo configuration, the human body acts as a moderator, producing scattered thermal and epithermal neutrons to which ^{6}Li has a higher response. The response of TLD albedo dosemeters drops for neutron energies higher than 100 keV, so field specific correction factors are required. These factors depend on the dosemeter itself and the characteristics of the field. TLD based albedo dosemeters have a useful dose range from ~ 0.1 mSv to 2 Sv.

Solid state nuclear track detectors are commonly used as neutron personal dosemeters and can measure doses from around 0.2 mSv to 200 mSv with a neutron energy range from ~150 keV to 15 MeV. The latent damage track caused by a neutron interaction in a suitable material is developed into visible pits by chemical or electro-chemical etching. An automated scanner is used to assess the number of such tracks and hence determine the dose. Poly allyl diglycol carbonate (PADC or Columbia Resin 39 (CR-39) or Tastrak) is the most widely used material.

2.1.2.2 Active whole body dosemeters

Active or electronic personal dosemeters (EPDs) give an instantaneous digital readout of accumulated dose and dose rate and enable the user to set dose/dose rate alarm levels. Electronic dosemeters are therefore very useful to warn the wearer when the dose rate or cumulative dose exceeds a certain value (e.g. for emergency services responding to a radiological emergency) and identify which particular procedures in a work process are more significant contributors to the total dose (Baptista *et al* 2015).

There are several manufacturers of EPDs. A variety of detectors are used including Geiger–Müller tubes, diodes, capacitors or direct ion storage devices. As with passive dosemeters, the multiple detectors can be placed behind different filter materials. These instruments will typically measure $H_p(10)$ between 1 μSv (or less) and 10 Sv. They are stable, accurate (e.g. within ±20%), reasonably rugged, and relatively insensitive to environmental changes in humidity, temperature and pressure, and electromagnetic fields.

Modern dosemeters that are designed as a direct replacement for the more traditional passive dosemeters (e.g. Instadose (Mirion Technologies, Smyrna, USA) and Verifi (Landauer, Glenwood, US, in development)) do not have a direct display so the user must obtain the dose by reading the device on a computer, perhaps via WiFi or bluetooth. This reduces the weight, size and cost. Linking the readout process directly to a web based programme results in the assessed doses going directly into the wearer's dose records. The wearer does not need to return the dosemeter for processing. Instadose 1 can measure doses from photons with energies above 15 keV from 0.01 mSv and is approved by the HSE in the UK.

The response down to low energy photons (20 keV) is particularly important for staff working in interventional radiology or with, for example, ^{125}I. Suitably selected EPDs have an energy and angular response that meets the International Electrotechnical Commission (IEC) standard (IEC 2010) but concern remains regarding the response of these instruments to pulsed radiation fields and very high dose rates such as may be found in the direct fields used in interventional radiology and radiotherapy. The i3 dosemeter (Raysafe, Billdal, Sweden) has been specifically designed for interventional radiology and is worn outside the lead apron. The dose and dose rate from several staff working in the interventional suite is wirelessly transferred to a real time display in the room, giving direct feedback so that staff can modify their behaviour to reduce their doses. The dosemeter is not offered by approved dosimetry services.

Bubble and direct ion storage detectors can be classed as active neutron dosemeters as they give an almost immediate readout. In addition truly active neutron dosemeters using silicon detectors are now commercially available.

2.1.3 Extremity and eye monitoring

TLDs are usually used for extremity and eye dose assessments because of their convenient size. The most commonly used phosphor is lithium fluoride. Extremity dosemeters measuring $H_p(0.07)$ need to be robust (because they may be placed on the hands carrying out manual work, etc) but also physically thin in order to avoid significant attenuation of any weakly penetrating radiation such as low energy β-radiation within the dosemeter. Extremity dosemeters therefore often consist of a thin uniform layer of phosphor sandwiched between a backing substance and a very thin layer of covering material. Placing the TL chip in a finger stall enables the dose at the very tip of the finger to be measured. Alternatively the chip can be placed in a plastic ring worn on the finger. The position of the dosemeter can have a significant impact on the value of the dose measured (see section 3.2.2). Photons in the energy range 15–1250 keV and β energies as low as 224 keV (E_{max}) can be measured. Doses as low as 0.15 mSv and up to 10 Sv are typically measured.

A variety of designs of eye dosemeters have recently entered the market (Behrens *et al* 2017). In order to measure $H_p(3)$, the dosemeter element normally used to measure $H_p(0.07)$ can be enclosed in a PVC pocket behind a plastic filter of suitable thickness. The resultant dosemeter can be attached to the (leaded) spectacles or inserted into a headband. Recently some designs allow the dosemeters to be attached to the inside (i.e. shielded part) of the leaded glasses. For interventional radiology, the dosemeter should be positioned at the side of the head nearest to the radiation source, close to the eye (Geber *et al* 2011). If the dosemeters are to be used for classified workers, it is important to check that the ADS is approved for eye dosimetry. The energy response and dose range for eye dosemeters are similar to those for extremity monitoring.

Electronic monitoring systems using a small probe attached in an appropriate position (e.g. ED3 system (John Caunt, Oxford) or the EDD-30 dosemeter (Unfors, Sweden)) can be used to give a continuous readout of dose rate to the hands, eyes or body. The real time dose information that is recorded can be downloaded for later analysis (Baptista *et al* 2015). This data coupled with observation of practices can be used to identify procedures which make significant contributions to the overall dose and help develop optimisation strategies to reduce exposure.

2.1.4 Technical standards and overall accuracy

International Standards stipulate the required performance characteristics for the various different types of dosemeters. IEC 62387-1 (CENELEC 2012) covers passive whole body and extremity dosemeters measuring $H_p(10)$ and $H_p(0.07)$ from x, γ and β radiations. For example, the variation in the relative response of a passive personal dosemeter, whilst measuring $H_p(10)$ over the photon energy range 65 keV to 1.25 MeV and for angles of incidence between 0° and ±60°, must be within –0.71 to 1.67. ISO 21909 (2005) covers passive neutron personal dosemeters. IEC 61526 (2010) covers direct reading (or active) whole body and extremity dosemeters for all types of radiations including neutrons.

The European Commission Technical Recommendations (2009) recommend that for the measurement of the operational quantities $H_p(10)$, $H_p(0.07)$ and $H_p(3)$ for single field components the combined uncertainty should be less than 30%. This applies at dose values equal to or greater than the relevant annual dose limit for members of the public in proportion to the wear period.

In practice the overall errors in the assessment of effective dose, extremity dose or eye dose in medical applications will be much larger than those mentioned above because of the errors associated with partial body exposure for staff wearing lead aprons, the very large dose gradients associated with extremity monitoring when handling radioactive material and the varying amount of protection offered by protective shields and clothing.

2.2 Internal monitoring

Evaluation of committed effective dose from internal monitoring is a complex process which requires:

- individual monitoring measurements to determine the activity, in Bq, of a particular radionuclide in a particular body organ or body component at a particular time;
- various models (published by the ICRP), which describe how the activity in different organs will vary over time after intake, to determine the actual initial intake of each radionuclide (also in Bq); and

- conversion of the intake into a committed effective dose using effective dose coefficients (ICRP 2012).

Individual monitoring can be achieved by direct *in vivo* measurement of activity in the body, excreta monitoring or air sampling with personal air samplers. Nose blow samples or nasal smears can give a useful preliminary measure of internal contamination following an incident.

In vivo measurement is only feasible for radionuclides which emit radiation that can escape from the body. ^{131}I, ^{125}I, ^{129}I, ^{137}Cs and ^{60}Co can be detected by comparatively simple equipment such as a single detector viewing the whole body or, for iodine, a smaller detector placed close to the thyroid. Excreta monitoring is required for radionuclides with no gamma emission. Routine programmes usually involve analysis of urine or faeces. Analysis of β- and α-emitters requires chemical separation before the appropriate measurement technique. Urine monitoring can be used to detect ^{125}I (liquid scintillation counting), ^{129}I (radiochemical separation and counting) and ^{131}I (γ-spectrometry). Sampling of airborne particulate materials can be carried out by drawing air through a filter using a calibrated air pump. The activity deposited on the filter, or a collection of filters, can be determined by radiochemical separation and high-sensitivity measurement techniques.

The ICRP consider it is generally inappropriate to formally assess doses with intake measurements when they are lower than 1 mSv in a year (ICRP 1997). The International Organisation for Standardisation (ISO) suggests that individual monitoring (e.g. *in vivo* monitoring, personal air samples) is carried out when the worker could be exposed to doses more than 30% of the limit by internal exposure (ISO 2006). Workplace monitoring such as workplace air sampling may be appropriate at lower doses. There is little data on internal doses to nuclear medicine workers. Airborne contamination levels resulting from ^{99m}Tc used for lung ventilation studies (e.g. Technegas or aerosols with DTPA) result is very small effective doses (López Medina *et al* 1999, Achey *et al* 2004). Internal monitoring has been carried out on nuclear medicine workers overseas who handle ^{131}I. Typical annual committed effective doses were of the order of 1 mSv with some individuals receiving higher values (Lucena *et al* 2007, Lee *et al* 2012, Krajewska and Krajewski 2012). These results were based on single measurements extrapolated for a full year assuming constant exposure levels throughout the year.

Thus, routine internal monitoring for nuclear medicine workers is unlikely to be justified, but occasional monitoring to check intakes of radioactive iodine may be necessary, especially in the event of an accidental spill or radiological incident.

Practical screening methods to identify workers most at risk have been proposed for different radionuclides. A calibrated surface contamination monitor placed on the thyroid can be used for radioiodine and daily measurements of ambient dose rate, $H^*(10)$, in front of the abdomen are recommended for shorter-lived radionuclides such as ^{99m}Tc (Baechler *et al* 2011). Such measurements are sufficient to confirm that a given threshold activity intake has not been exceeded (rather than an accurate determination of effective dose).

The European Commission has published “Technical Recommendations for Monitoring Individuals for Occupational Intakes of Radionuclides” (European Commission 2018). The report explains the principles of internal dosimetry and individual monitoring, and presents comprehensive, detailed, authoritative and internally-consistent guidance and recommendations on practice.

2.3 Use of electronic monitors

The personal dosimetry measurements discussed in the earlier sections of this chapter have been designed to demonstrate compliance with dose limits and for comparison with investigation levels. The personal dose monitors worn by staff integrate the doses over a variety of procedures performed during a prescribed monitoring period. Over time, the impact of optimisation should be apparent through lower occupational doses for comparable workloads. However, this form of dosimetry does not provide information on doses received for different activities and types of procedure. Optimisation monitoring needs to identify where and when higher doses are received and evaluate the effectiveness of actions and controls to reduce staff doses. As discussed in section 2.1.2.2, many EPDs are able to provide data on cumulative doses and dose rates over time that can be used to identify the procedures in a work process that make more significant contributions to the total dose. Many current models provide dose and dose rate measurements over short time increments, which can be downloaded into spreadsheets for analysis. Periods when dose rate levels are high can then be identified and associated with the particular actions which were responsible (sections 3.5 and 5.5.2). EPDs can also be helpful in monitoring personnel with multiple employers, enabling the dose received to be recorded at the end of a session.

EPDs for use in optimisation do not necessarily need to fulfil the more stringent requirements for measurement of personnel doses to be included in statutory dose records. A variety of such EPDs are available and the performance of instruments for interventional radiology and cardiology from several manufacturers have been compared and found to be within ±20%–30% (Chiriotti *et al* 2011, Struelens *et al* 2011), although Struelens *et al* report that the median value of the measured doses was 20% lower than values recorded by passive dosemeters. The ability to record low energy photons and β-radiation must be considered in judging the suitability of a particular model for application in a specific area. Although these dosemeters are generally robust and perform well under a range of conditions, problems have been reported when measuring pulsed fields and this can be an issue for monitoring in interventional suites (Ankerhold *et al* 2009, Ginjaume 2011).

Advances in power management and wireless transmission of signals from EPDs to a base station have facilitated data collection and transmission. Systems are now available for use in interventional suits with EPDs that have real-time wireless data transmission so that dose rates and doses can be viewed on a screen by the operator and staff present in the room, giving direct feedback on dose levels during the procedure (Sandblom *et al* 2013). As technological developments continue, new opportunities for dose assessment can enable a better appreciation of staff exposure and should be used in improving the understanding and education of staff. However, care should be taken that staff are not distracted from their clinical work by constantly looking at the dose monitor screen.

References

Achey B, Miller K, Erdman M and King S 2004 Potential dose to nuclear medicine technologists from 99m-Tc-DTPA aerosol lung studies *Health Phys.* **86** 85–7

Ankerhold U, Hupe O and Ambrosi P 2009 Deficiencies of active radiation protection dosemeters in pulsed fields *Radiat. Prot. Dosimetry* **135** 149–53

Baptista M, Figueira C, Teles P, Cardoso G, Zankl M and Vaz P 2015 *Radiat. Prot. Dosimetry* **165** 304–9

Baechler S, Stritt N and Bochud F O 2011 Individual monitoring of internal exposure for nuclear medicine workers in Switzerland *Radiat. Prot. Dosimetry* **144** 464–7

Behrens R, Hupe O, Busch F, Denk J, Engelhardt K, Günther K, Hödlmoser H, Jordan M and Strohumaier J 2017 Intercomparison of eye lens dosemeters *Radiat. Prot. Dosimetry* **174** 6–12

CENELEC 2012 *IEC 62387-1. Radiation Protection Instrumentation - Passive Integrating Dosimetry Systems for Environmental and Personal Monitoring - Part 1: General Characteristics and Performance Requirements* (Brussels, Belgium: CENELEC)

Chiriotti S, Ginjaume M, Vañó E, Sanchez R, Fernandez J M, Duch M A and Sempau J 2011 Performance of several active personal dosemeters in interventional radiology and cardiology *Radiat. Meas.* **46** 1266–70

European Commission 2009 *Radiation Protection No.160. Technical Recommendations for Monitoring Individuals Occupationally Exposed to External Radiation* (Luxembourg: Office for Official Publications of European Union) https://ec.europa.eu/energy/sites/ener/files/documents/160.pdf

European Commission 2018 Radiation Protection No.188. Technical Recommendations for monitoring individuals for occupational intakes of radionuclides. Office for Official Publications of European Union, Luxembourg. https://ec.europa.eu/energy/sites/ener/files/rp_188.pdf (accessed 20/9/18)

Geber T, Gunnarsson M and Mattson S 2011 Eye lens dosimetry for interventional procedures - relationship between the absorbed dose to the lens and dose at measurement positions *Radiat. Meas.* **46** 1248–51

Ginjaume M 2011 Performance and approval procedures for active personal dosemeters *Radiat. Prot. Dosimetry* **144** 144–9

ICRP 1997 Individual monitoring for internal exposure of workers. ICRP publication 78 *Ann. ICRP* **27**

ICRP 2007 The 2007 recommendations of the International Commission on Radiological Protection. ICRP Publication 103 *Ann. ICRP* **37** 1–332

ICRP 2012 Compendium of dose coefficients based on ICRP Publication 60. ICRP Publication 119 *Ann. ICRP* **41** 1–130

IEC 2010 *IEC 61526 Radiation Protection Instrumentation - Measurement of Hp(10) and Hp (0.07) for X, Gamma, Neutron, and Beta Radiations - Direct Reading Personal Dose Equivalent Meters* (Geneva, Switzerland: IEC)

ISO 2005 *ISO 21909 Passive Personal Neutron Dosemeters - Performance and Test Requirements* (London, UK: British Standards Institution)

ISO 2006 *ISO 20553 Radiation Protection - Monitoring of Workers Occupationally Exposed to a Risk of Internal Contamination with Radioactive Material* (London, UK: British Standards Institution)

Krajewska G and Krajewski P 2012 Evaluation of internal exposure of nuclear medicine staff working with radioiodine and technetium *IRPA13 (Glasgow)* (abstract) http://www.irpa.net/members/IRPA13-abstract-USB-FINAL.pdf (accessed 21/3/16)

Lee J I, Kim B H, Jeong K H, Lee S H and Lim K 2012 The assessment of internal doses for Korean nuclear medicine workers based on thyroid bioassay measurement of 131I *IRPA13 (Glasgow)* (abstract) http://www.irpa.net/members/IRPA13-abstract-USB-FINAL.pdf (accessed 21/6/16)

Lucena E A, Rebelo A M O, Araujo F, Sousa W O, Dantas A L A, Dantas B M and Corbo R 2007 Evaluation of internal exposure of nuclear medicine staff through in-vivo and in-vitro bioassay techniques *Radiat. Prot. Dosimetry* **127** 465–8

López Medina A, Miñano J A, Terrón J A, Bullejos J A, Guerrero T, Arroyo T, Ramírez A and Llamas J M 1999 Calculation of airborne radioactivity in Technegas lung ventilation unit *Nucl. Med. Commun.* **20** 1141–5

Sandblom V, Mai T, Almén A, Rystedt H, Cederblad A, Båth M and Lundh C 2013 Evaluation of the impact of a system for real-time visualisation of occupational radiation dose rate during fluoroscopically guided procedures *J. Radiol. Prot.* **33** 693–702

Struelens L, Carinou E, Clairand I, Donadille L, Ginjaume M, Koukorava C, Krim S, Mol H, Sans-Merce M and Vanhavere F 2011 Use of active personal dosemeters in interventional radiology and cardiology: Tests in hospitals - ORAMED project *Radiat. Meas.* **46** 1258–61

IOP Publishing

Colin John Martin, David Temperton, Anthony Hughes and Thomas Jupp

Chapter 3

Nuclear medicine

3.1 Introduction

In nuclear medicine (NM) staff are exposed to radiation emitted directly by radiopharmaceuticals contained within vials and syringes, and sealed sources used for quality control. There is also exposure from proximity to patients containing radioactivity and potentially from handling radioactive waste. The handling of radiopharmaceuticals can lead to significant exposure of the hands as well as the whole body. Protection of the body when handling radiopharmaceuticals is provided through shielding of vials and syringes, and bench top shields, while protection when dealing with patients is, in most cases, achieved through maintaining distance and limiting the time spent in close contact. Manipulation of unsealed and sealed sources during the preparation and administration of radiopharmaceuticals can lead to high doses to the hands as well as doses to the eyes. Measures to protect the fingers are particularly important and recent studies have revealed poor implementation of protection measures and limited monitoring of fingers in many parts of Europe (Donadile 2008, Carnicer *et al* 2011). Typically occupational exposure of NM personnel is monitored using a single dosemeter worn on the trunk to assess the whole body dose and either finger stalls worn at the tip or ring dosemeters worn at the base of the finger to assess finger doses.

There may be high dose gradients across the hands of NM workers (Whitby and Martin 2003, ICRP 2008, Vanhavere *et al* 2012), especially when handling β^- or β^+ sources (Chruscielewski *et al* 2002). This can complicate routine monitoring since the maximum dose, which is usually at the fingertips, is systematically underestimated by ring dosemeters.

Implementation of the European Directive (EC 2013) requires more attention and vigilance in estimation of eye and extremity doses. But to ensure effective monitoring and compliance with the dose limits the convenience of dosemeters currently

doi:10.1088/2053-2563/aaddb6ch3

Table 3.1. Guidance on the need for dose monitoring at different dose levels (based on information reported in Martin and Whitby 2003).

Tissue and dose quantity	Annual dose to tissue[c] (mSv)	Monthly dose to tissue (mSv)	Dose monitoring strategy
Body $H_p(10)$	> 1 (6)[a]	> 0.1	Monitoring at chest, waist or collar depending on exposure situation.
Eyes $H_p(3)$	> 6 (15)[a]	> 0.5	Regular monitoring with collar or head dosemeter is required.
Extremities $H_p(0.07)$	20–50[b]	2–4[b]	Initial monitoring to establish dose levels with ring dosemeters or finger stalls.
Extremities $H_p(0.07)$	> 50[b] (150)[a]	> 4[b]	Regular monitoring with finger stalls recommended. Establish hand ratios to correct dose, if ring dosemeter used.

[a] Dose level at which classification is required in brackets.
[b] Dose level evaluated by finger stalls, or application of locally assessed ratio or factor of 6 to measurements with a ring dosemeter.
[c] This should include doses from routine work and potential accidents.

employed for the whole body and fingers needs to be maintained. General guidance about the dose levels at which monitoring should be performed is given in table 3.1. More detailed recommendations and discussion of the issues involved are given in the text for individual areas of application.

3.2 Diagnostic nuclear medicine with γ-emitting radionuclides

3.2.1 Whole body dosimetry

In order to gain an appreciation of how doses are built up from the many different tasks undertaken in a NM department, the major contributions to NM technologists' doses from a study by Smart (2004) are listed in table 3.2 for a 900 MBq bone scan. The steady accumulation of dose from routine tasks throughout the day can result in NM staff receiving annual doses to the body of several millisieverts. Doses from the majority of routine tasks make small contributions, but some individual procedures such as cardiac scanning can give rather larger doses and need to be given special consideration. Although the distribution of dose is likely to be relatively uniform, there will be some variations. For example Lundberg *et al* (2002) found the average waist and collar dose readings for a NM technologist were similar for the majority of tasks, but for injections the collar readings were 1.7 times higher.

Specific attention should be applied to monitoring of pregnant staff. Potential body doses are at the level for which a pregnant worker could need to modify their work in order to minimise the dose to the foetus. The most effective strategy might be for a technologist who becomes pregnant to be assigned patients requiring little contact or assistance and to avoid administering injections, but the precautions and

Table 3.2. Examples of body doses that might be received from different tasks linked to a bone scan of a patient injected with 900 MBq (Smart 2004).

Task	Average dose	Potential dose for challenging patient
Transferring incapacitated patients from the imaging table to a hospital trolley	0.5 μSv	1.6 μSv
Injection using tungsten syringe shields	0.6 μSv	1.6 μSv
Positioning patient for gamma camera imaging	0.4 μSv	

dose monitoring would need to be tailored to the individual and addressed in a risk assessment that included requirements for dose monitoring (see section 7.3).

> NM staff are recommended to wear a body dosemeter. For the majority of imaging tasks radiation exposure is comparatively uniform, so dosemeters can be worn either at waist, chest, or collar level. However, it is appropriate to review the exposure and risks for the tasks performed when recommending positions for dosemeters.
>
> The waist is the most suitable location for monitoring pregnant staff. For staff working with ^{99m}Tc, a dose of 1.3 mSv to the surface of the abdomen equates to a foetal dose of about 1 mSv.

3.2.2 Extremity dose monitoring

3.2.2.1 Protecting the hands

NM services involve both the preparation and administration to patients of radiopharmaceuticals. Some of the actions involved in the preparation and administration stage are similar but entail handling different activities and volumes of radioactivity. The preparation stage is usually carried out in a Radiopharmacy, which may handle between 20 GBq and 100 GBq per day. The tips of the fingers receive the highest doses when manipulating syringes and vials. The dose received is dependent upon the technique used to dispense the radiopharmaceuticals and the amount of shielding used. Use of a vial shield rather than a simple lead pot helps to reduce finger dose during dispensing in radiopharmacies, and syringe shields are recommended for radiopharmacy and NM procedures, namely drawing up and performing injections, as they reduce finger doses substantially. Poor visibility of the liquid in the syringe can be an issue, but the reduction in dose is such as to more than justify their use for the majority of applications. Table 3.3 shows relative values of doses to the fingertip in six radiopharmacies that used different dispensing techniques and demonstrates that effective shielding can reduce the finger dose by 75–85% (see also section 3.5) (Whitby and Martin 2003, 2005).

Table 3.3. Data showing influence from use of syringe shields and dispensing technique on finger dose from six UK radiopharmacies. Some of the drawing-up techniques are illustrated in figure 3.1 (data from Whitby and Martin 2005).

Centre	Dispensing technique (see section 3.2.2.2)	Use of syringe shields	Absorbed dose (μGy GBq^{-1})
A	Full inversion	Complete	3.8
		Partial	13.0
		None	35.0
B	45° table top	Partial	6.6
C	V-technique	Complete	4.2
D	Modified V-technique	Limited use	21.3
		Mostly	4.9
E	Jig	Complete	4.4
F	Modified V-technique	Variable	6.7–39.3

3.2.2.2 Distribution of dose across the hand

The study by Whitby and Martin (2005) showed that the tips of the index finger and thumb of the dominant hand holding the syringe tended to receive the largest doses. However, there were several variants of the technique for withdrawing activity from a vial into a syringe, some of which are illustrated in figure 3.1. These influence the finger on the non-dominant hand receiving the higher dose. For example with the inversion technique (figure 3.1(a)) the wrist or little finger of the non-dominant hand might receive a higher dose, whereas with the V-technique (figure 3.1(c)) the index finger of the non-dominant hand was more exposed. The technique can be more influential for dose distributions in NM, where activity is being withdrawn from a vial held in a lead pot. Here the vial is supported by the syringe needle, but the neck may protrude from the shield and occasionally a finger may touch the cap of the vial to keep it in place. This would be the little finger of the non-dominant hand in figure 3.1(a), but the index finger in figure 3.1(c). Therefore, careful observation of staff practices is required when making recommendations about dosemeter placement. Similar measurements have been made in 30 NM departments from six European countries as part of the ORAMED project, and these showed higher finger doses ranging from about 20 to 800 μSv per GBq for both preparation and administration of ^{99m}Tc (Struelens *et al* 2010, Carnicer *et al* 2011, Vanhavere *et al* 2012). In these studies the tips of the index finger and thumb on the two hands were found to receive the highest doses, and the hand receiving the higher dose depended on the technique of the individual.

3.2.2.3 General rules for determining choice of finger dosemeter

If annual extremity doses are a few tens of mSv, the risk assessment is likely to conclude that ring dosemeters worn at the base of the finger or at the position of the second phalanx can be used for routine monitoring, but an attempt should be made to determine how the dose at the position used relates to that at the most exposed

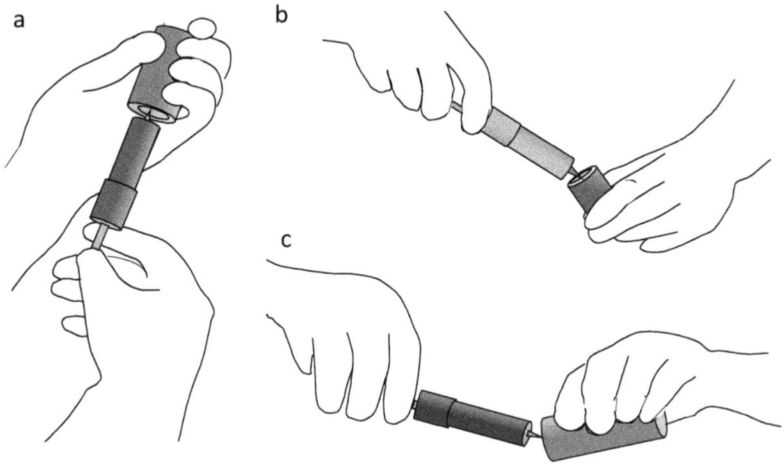

Figure 3.1. Hand positions used in different techniques for withdrawing radiopharmaceutical from a vial using a syringe: (a) full inversion, (b) 45° table top using a syringe with a longer needle, and (c) V technique (Whitby and Martin 2005). N.B. Such manipulations should always be undertaken over a drip tray behind a protective barrier.

region, which is frequently the tip of the index finger. If the dose to a fingertip is likely to approach 100 mSv and the dose distribution across the hand is not uniform, then finger stalls should be worn in order to measure the dose at the position considered to receive the highest dose (table 3.1). For small sources such as syringes and vials containing radioactive liquid, the magnitude of the dose gradient is determined by the proximity of the fingertip to the unshielded source. When choosing the type of finger monitoring to be adopted, it may be helpful to apply some basic principles and suggestions are given below:

1. if fingertips are within 5 cm of the unshielded source for the majority of the time, or may come into direct contact with the source, even for relatively short periods, finger stalls provide the better option;
2. if the radiation field exposing the hand is uniform or the hand is always more than 5 cm from the source, then ring dosemeters can be used;
3. if fingertips are within 5 cm of the source, but the exposures are all received from manipulations that are repeated many times, and the doses do not approach 100 mSv, then ring dosemeters provide a possible option and an adjustment factor could be applied by the radiation protection service in order to derive an estimate of the dose to the fingertip.

The third option might be appropriate for staff for whom the wearing of dosemeters may not be acceptable, because of the effect on touch sensation and the ability to make fine adjustments. However, such an approach would not be appropriate for staff working with therapy radionuclides emitting positron or β-radiations, or where staff members undertake several different roles within a department in imaging, therapy, and potentially a radiopharmacy. Adjustment or correction factors cannot be applied to the dose records by an ADS until the process has been included in the ADS's statement of service and approved by HSE. The employer must have a system in place to confirm that the dosemeter is always worn in the designated position and the factors applied must be validated at regular intervals to ensure that they remain suitable for each specific wearer.

3.2.2.4 Correction factors for application to ring dosemeter measurements

Ring dosemeters may underestimate doses to the fingertips significantly, because of the large dose gradients close to the source. This will be especially true if sources are not well shielded. Therefore, if ring dosemeters are used, techniques are required that allow fingertip dose to be estimated from that recorded by a ring dosemeter. There have been numerous studies of dose distribution across the hands during the preparation and administration of ^{99m}Tc labelled pharmaceuticals in UK NM departments (Harding *et al* 1985, Williams *et al* 1987, Batchelor *et al* 1991, Montgomery *et al* 1999, Dhanse *et al* 2000, Whitby and Martin 2003, 2005) and ^{18}F (Morton *et al* 2006). In the UK, syringe shields will normally be used during both the preparation and administration stages, and a butterfly cannula is frequently inserted into a vein, prior to radiopharmaceutical administration to avoid the need to use the fingertip to guide the needle when it contains radiopharmaceutical (Martin 2016). Provided the tubing is not too long (e.g. 95 mm) significant dose reductions can be achieved (Batchelor *et al* 1991). The UK studies found ratios between doses to the tip and base of the index finger were in the range 2–4. The ORAMED results support a scaling factor of three between the base and tip of the index finger for both the dominant and non-dominant hands, but also demonstrate that the location of maximum dose is not necessarily the tip of the index finger (Vanhavere *et al* 2012). These data therefore suggest a factor of six to scale doses measured at the base of the index finger to the maximum dose. However, it must be borne in mind that practices in this study were variable and examples of poor practice included injections without syringe shields and staff using the tip of the finger of the non-dominant hand to guide the needle during insertion into a vein for injection. If a ring dosemeter is worn on the middle phalanx closer to the tip and protection had been optimised, a factor of two might be more appropriate (Martin 2016).

3.2.2.5 Methodology for assessing ratio of fingertip to ring dosemeter reading

A suggested methodology for determining scaling factors is set out in the following. Similar methods could be applied to PET procedures using ^{18}F, but here more caution should be exercised in interpretation of results and checks made with finger stalls at more regular intervals.

- Observe the configuration of both hands used by each staff member when drawing up radiopharmaceutical and giving injections, and discuss the technique with individual staff. Determine whether any fingertips are likely to be more exposed than other parts of the hand. If they are, decide which finger(s) on either hand would be the most likely to be involved. This will probably be the index fingers for most individuals.

- Each operator should wear dosemeters on the tips of the fingers identified on either hand. Ring dosemeters should also be worn at the bases of the fingers chosen with the dosemeter facing inwards towards the palm to allow comparisons. Both dosemeters should measure the same dose quantity, preferably $H_p(0.07)$ (Carnicer *et al* 2011). N.B. It is essential to use dosemeters recording $H_p(0.07)$ for PET or β-emitting radionuclides.
- Monitoring at the chosen sites should be conducted for a minimum of three consecutive sets of dosemeter readings to provide an assessment of the maximum dose and appropriate scaling factors for the ring dosemeters. If dose ratios are variable the trial should be continued for a longer period.
- If the dose levels are variable, the better option for deriving a factor is to sum the doses from each position and take the ratio, rather than deriving a ratio for each period.
- Once a stable scaling factor has been determined routine monitoring can then be performed using the ring dosemeters alone.
- Monitoring using finger stalls should be repeated annually for ^{99m}Tc applications and more frequently for PET radionuclides, or whenever anomalous results are obtained with the ring dosemeters.

The application of correction factors for non-classified staff by the employer should allow estimation of doses to the fingertips, in order to determine whether more extensive monitoring is appropriate and the possible need for classification. The value recorded by a ring dosemeter can be recorded as the dose to the whole hand. But where there is a risk of the dose to a fingertip approaching 100 mSv, finger stalls are recommended.

3.2.2.6 Recommendations for finger dose monitoring

Monitoring of doses to the hands is recommended for all radiopharmacy staff preparing ^{99m}Tc radiopharmaceuticals and for staff who administer imaging radiopharmaceuticals regularly. Use of finger stalls to measure doses to the tips of the index fingers on both hands is recommended wherever there is a risk of the fingertip dose approaching 100 mSv, provided this does not adversely affect manipulations (see table 3.1).

If ring dosemeters are worn, rather than finger stalls, it is recommended that the detector element is on the palmar side of the index fingers on either hand and a correction factor is applied to estimate doses to the fingertips. Measurements at the tip should be carried out for a trial period in order to establish a ratio between doses to the tip and the position where the dosemeter is worn that can subsequently be applied routinely. Adjustment or correction factors cannot be applied to the dose records by an ADS until the process has been included in the ADS's statement of service and approved by HSE. The employer must have a system in place to confirm that the dosemeter is always worn in the designated position and the factors applied must be validated at regular intervals to ensure that they remain suitable for each specific wearer. For non-classified workers, where a reliable correction factor is not available from local measurements, application of a correction factor of six is recommended for measurements from dosemeters worn at the base of the finger to provide a conservative estimate of the maximum dose. Alternatively a ring dosemeter may be worn on the second phalanx, closer to the tip, and a factor of two applied (Martin 2016). The correction factor could be applied by the employer.

3.2.3 Eye dose monitoring

Doses to the eyes of NM staff are unlikely to approach the revised dose limit (EC 2013, IRR17), so the need to monitor eye dose with a dedicated eye dosemeter would only be expected in specific circumstances (Magee 2015, Dabin *et al* 2016, Wrzesień *et al* 2018). A recent European survey by members of WG12 of EURADOS (Eleftheria *et al* 2014) reported that eye doses are not measured or estimated routinely in NM in Europe as a whole. Since exposure for the majority of tasks is relatively uniform, and personal protective equipment is not worn generally, dose data available from dosimetry records should provide a reasonable basis for evaluating potential exposure levels for the majority of NM staff, provided appropriate techniques and shielding are employed (Prince and Murby 2014, Magee 2015). However, there may be certain situations where a body shield is used, for example whilst setting up a patient on a PET scanner or preparing radiopharmaceuticals in a larger radiopharmacy, when the dosemeter is behind the shield, but the head and shoulders are not protected. In these circumstances specific monitoring of the eye lens dose may be required at least for an initial period. However, the lowering of the eye dose limit is unlikely to require any additional NM personnel to become classified persons (Category A workers).

As described earlier, different dose quantities are used for monitoring, depending on the depth of the sensitive tissue. The personal dose equivalent $H_p(10)$ is used for whole body measurements, $H_p(3)$ for eye lens, and $H_p(0.07)$ for skin and extremities. ^{99m}Tc is the predominant radionuclide used in NM and 88% of the atoms decay through emission of a 140 keV γ-ray. However, the other 12% of atoms decay through internal conversion and emit a low energy electron that is heavily absorbed in superficial tissues and this contributes to the $H_p(0.07)$ reading. Since there are limited $H_p(3)$ dosemeters in use at the present time, a comparison of measurements for $H_p(10)$, $H_p(3)$ and $H_p(0.07)$ for 11 staff in a NM department (Kopeć *et al* 2011) is informative (table 3.4). The results suggest that a reasonable estimate of the $H_p(3)$ eye lens dose can be derived from the $H_p(10)$ dose measured from conventional whole body dosemeters worn on the chest.

A whole body dosemeter worn on the chest or collar should provide sufficient data for estimating dose to the lens of the eye for the majority of NM staff. If a measure of $H_p(3)$ is not available, then it is recommended that the $H_p(10)$ dose is used for ^{99m}Tc.

Table 3.4. Ratios of $H_p(3)$ and $H_p(10)$ for NM staff working with ^{99m}Tc (Kopeć *et al* 2011).

Staff group	Ratio $H_p(3)/H_p(10)$	Ratio $H_p(3)/H_p(0.07)$
Technical staff (preparation)	1.1 ± 0.2	0.09 ± 0.02
Technical staff (operator)	0.7 ± 0.1	0.14 ± 0.013
Nurses	0.7 ± 0.2	0.08 ± 0.01
Physician	0.8 ± 0.3	0.09 ± 0.03

3.3 ^{18}FDG-positron emission tomography (PET) imaging

3.3.1 Whole body dosimetry

Whole-body doses have been reported for PET imaging using either TLD or OSL dosemeters worn on the trunk (Pant *et al* 2006, Kopeć *et al* 2011, Guillet *et al* 2005, Roberts *et al* 2005, Leide-Svegborn 2012, Seierstad *et al* 2007, Benatar *et al* 2000). These published studies have recorded total whole body dose factors in the range 1–25 μSv per GBq activity administered. The largest component of this dose is received during the injection (Guillet *et al* 2005, Peet *et al* 2012). This is to be expected given that the external dose rates from patients injected with ^{18}FDG are much higher than in conventional nuclear medicine and are typically 45 μSv h^{-1} at 1 m (Benatar *et al* 2000, Leide-Svegborn 2012, Peet *et al* 2012). Published whole body dose values for PET staff are in the range 4–10 μSv per patient (350 MBq per injection). So for a workload of 500 patients per annum the expected effective dose will be 2–5 mSv and these workers may need to be classified in order to take account of potential doses from incidents involving periods of higher exposure levels.

> Monthly whole body monitoring with a dosemeter worn at the waist or chest is recommended as standard for staff involved in PET imaging.

3.3.2 Finger dose monitoring

3.3.2.1 Extremity dosemeters

Extremity doses should be measured in terms of $H_p(0.07)$. Many dosemeters underestimate skin doses from ^{18}F, ^{68}Ga, and other radionuclides emitting positrons or β-particles. The radiation interactions are averaged out over the volume of a dosemeter, so if the dosemeter is too thick, the measurement will not reflect high doses occurring at the surface that result from β-radiation. This is less of a problem for more penetrating γ-radiations.

3.3.2.2 PET scanning

Finger monitoring during preparation and administration of ^{18}FDG-PET doses was addressed by the ORAMED project (Vanhavere *et al* 2012). $H_p(0.07)$ was monitored at 22 different positions across both hands for 30 workers from six European countries. The maximum recorded dose normalised to the activity handled ranged from 0.10–4.4 mSv GBq^{-1} for preparation and 0.14–4.1 mSv GBq^{-1} for administration. The very wide range of doses recorded reflects different techniques, training and use of protective equipment at the test sites. It is acknowledged in the ORAMED report that 43% of the preparation measurements were performed without a syringe shield. Wrzesień and Napolska (2015) report fingertip doses of 0.2 mSv GBq^{-1} for preparation using an automated dispensing system, and 0.5 mSv GBq^{-1} for administration. Although syringe shields were used for the injections in this study, the tips of the index finger and thumb lay in the unprotection region near the end of the shield and staff giving injections received an annual dose over 100 mSv.

Using the largest of these recorded dose values an operator preparing and administering patient doses of 350 MBq would exceed 150 mSv after only 50 procedures and would breach the dose limit after 168 (this corresponds to just 33 working days in which five patient doses of 350 MBq are prepared and administered). Worryingly, even using the median values for the maximum normalised hand dose (0.83 and 0.64 mSv GBq^{-1} for preparation and administration respectively) the annual dose limit would be exceeded by staff preparing and administering 972 doses of 350 MBq (equivalent to 194 working days with five patient administrations). Although such doses are unlikely to be realised in the UK where shielding is routinely used during source manipulation, they emphasise the necessity for both routine finger monitoring and the routine use of syringe guards.

The ORAMED project also sought to identify the location of the maximum skin dose. They found that during preparation of ^{18}FDG injections the maximum dose was recorded at the tip of the index finger of the non-dominant hand in approximately one-quarter of all measurements. However, the recorded values demonstrate wide variations making it difficult to predict the actual location for any individual worker. During administration, the tips of the index finger on each hand were equally likely to receive the highest dose in two-thirds of the dose measurements. In a similar study conducted on 18 workers during the preparation of ^{18}FDG injections the tip of the index finger on the dominant hand was found to be the location of the maximum dose for 39% of the workers (Struelens *et al* 2010). The same authors also found the tip of the index finger and thumb of the dominant hand to be the most common site of the measured maximum dose during the administration of ^{18}FDG (15 workers), but concluded that the measurements were so variable that it was not possible to predict which hand would receive the maximum dose for individual workers.

Covens *et al* (2010) measured $H_p(0.07)$ at 36 positions across the hands of eight technologists during the preparation and administration of ^{18}FDG before and after the introduction of an automated dispensing/injection system. Maximum hand doses decreased from 0.3 to 0.01 mSv GBq^{-1} during dose preparation and from 0.4 to 0.015 mSv GBq^{-1} during administration. For manual procedures, the location of the maximum dose was at the tip of the left ring finger during preparation and at the tip of the left index finger during administration. As well as lowering the maximum dose received, automated procedures tended to result in more even dose distribution across both hands. Other studies have quoted total finger dose factors of 0.3–2.7 mSv GBq^{-1} (preparation and administration combined), but with no obvious pattern in the location of the maximum dose (Linemann *et al* 2000, Guillet *et al* 2005, Tandon *et al* 2007, Demir *et al* 2010, Kopeć *et al* 2011, Leide-Svegborn 2012, Antic *et al* 2014, Wrzesień and Napolska 2015).

Given the potential for high extremity doses continuous monitoring is essential for staff handling ^{18}F. The method depends on the techniques used and how they influence the positions of the fingers in relation to the source. If the fingertips are likely to receive the highest doses, then finger stalls worn on the more exposed fingers are recommended. The ORAMED project concludes that routine finger monitoring could be performed using a ring TLD at the base of the index finger of the non-dominant hand and that the recorded dose should be scaled by a factor of six to represent the maximum hand dose. However, there was a large variation in doses and the ×6 multiplier was a median, so the adoption of a single scaling factor is questionable. The variation in dose distribution across the hands and the uncertainties in both the location and magnitude of the maximum dose prevent an accurate assessment of dose using a ring dosemeter, if the fingers come within about 5 cm of the source. If an automated dispensing/injection system is used and the dose distribution is more uniform, then ring dosemeters might be suitable.

> For staff working with PET imaging, monitoring with a finger stall is recommended, if the fingers come within 5 cm of the source. Ring dosemeters worn at the bases of the index fingers on both hands may be used, where the dose distribution across the hand is more uniform, but to ensure that the doses to the more exposed fingers do not approach the dose limit scaling factors should be applied to estimate the doses to the fingertips, which should be assessed individually for each worker (see section 3.2.2.5). If a reliable scaling factor is not available from local measurements, a correction factor of six could be adopted to give some indication of the likely range for the maximum dose. As stated in section 3.1 adjustment or correction factors cannot be applied to the dose records by an ADS until the process has been included in the ADS's statement of service and approved by HSE. The employer must have a system in place to confirm that the dosemeter is always worn in the designated position and the factors applied must be validated at regular intervals to ensure that they remain suitable for each specific wearer. Therefore, the wearing of finger stall dosemeters is recommended for classified radiation workers handling PET radionuclides.

3.3.2.3 Cyclotron maintenance

Another group of staff, who may receive high finger doses are engineers who carry out maintenance of cyclotrons. Activation products in the target can deliver high dose rates. Most will decay within 24 h, so maintenance can be put back 24 h, but there are some longer half-life activation products such as ^{60}Co, ^{54}Mn, ^{22}Na, and ^{65}Zn, depending on the material that can still deliver significant doses. The dose rate adjacent to the target should always be measured prior to any work being carried out in order to assess the level of hazard, and practice should be based on risk assessment of the procedure. The main hazard is from short-range beta particles, if manipulation of the target is required. Finger contact should be avoided and appropriate handling tools used wherever possible, but if, during some parts of maintenance, fingertips could come close to the target, then finger stalls are recommended. The potential exposure should be evaluated in a risk assessment prior to commencement of this type of work.

3.3.3 Eye dose monitoring

Eye doses during ^{18}F manipulations have been measured using TLDs calibrated to measure $H_p(3)$ (Bilski *et al* 2011). Walsh *et al* (2014) measured eye dose for a number of different workers using a specially designed dosemeter placed adjacent to the left eye using an elasticated headband (EYE-D^{TM}, Bilski *et al* 2011). Of the staff groups included in the study, radiographers performing the administrations received the highest doses. However, in all cases the estimated annual eye dose was $\leqslant$ 2 mSv. When compared to $H_p(10)$ measured using an EPD worn at waist level $H_p(3)$ was found to be up to 50% greater in magnitude. Kopeć *et al* (2011) also measured $H_p(10)$ at the chest and $H_p(3)$ using the EYE-D^{TM} dosemeter for staff performing different tasks at two PET centres. They measured eye dose factors of ~2 μSv GBq^{-1} during preparation and ~3 μSv GBq^{-1} during administration and showed that $H_p(3)$ could be estimated from the measured $H_p(10)$ value. Based on their data, the expected eye dose for a member of staff preparing and administering 1000 doses of 350 MBq in a year would be approximately 2 mSv. However, Wrzesień *et al* (2018) have reported eye doses up to 0.2 mSv per day from the preparation and administration of ^{18}F FDG, and this would exceed the dose limit for the eye. Therefore although the majority of published eye dose values suggest that it is unlikely that additional routine monitoring of doses to the eyes will be required for PET workers, this may be necessary if protection is not optimised. At the moment, there is insufficient information to determine the optimum location for wearing of a dedicated eye dosemeter, but there is some indication that the dose distribution across the head is not uniform (Kopeć *et al* 2011).

> Measurements from a routine whole body dosemeter worn on the chest should give a measure of probable eye dose levels and an indication of whether dedicated measurements of eye dose will be required (see table 3.1).

3.4. Therapy with unsealed radionuclides

3.4.1 Whole body dosimetry

Although the majority of nuclear medicine personnel are not classified radiation workers they are issued with a passive dosemeter routinely for monitoring whole body doses and there is no evidence to suggest that additional monitoring is required for therapeutic procedures. An electronic personal dosemeter (EPD) can be worn for the purposes of risk assessment or for the partition of doses from different procedures, however, the energy response of the dosemeter must be appropriate for the emitted radiation and where β-emitters are concerned this could include a wide spectrum of bremsstrahlung x-rays (Jodal 2009). There is little published data on the variation of external dose rates across the body in different scenarios, so a suitable location for any dosemeter will need to be verified locally during risk assessment.

> Body dosemeters worn by NM staff are sufficient for recording whole body doses for therapeutic procedures.

3.4.2 Finger monitoring

Since radionuclides used for therapy emit particle radiations which are capable of delivering significant doses, consideration of potential finger doses and protection is extremely important. The requirements for extremity monitoring are related to the radionuclide in question. As for PET procedures, selection of a dosemeter recording $H_p(0.07)$ is required to avoid underestimating the dose to the skin surface (see section 3.3.2.1).

3.4.2.1 ^{90}Y therapy

To date, the largest survey of extremity doses for therapy involving ^{90}Y has been conducted by the ORAMED project (Vanhavere *et al* 2012). $H_p(0.07)$ was measured at 11 separate positions across each hand using LiF TLDs for two therapeutic procedures; RIT ^{90}Y-Zevalin® and ^{90}Y-DOTATOC. Separate measurements were made during preparation and administration of the treatment and the maximum dose at each location was recorded and normalised to the activity handled. Both therapies exhibited a wide range of maximum doses for preparation and administration, see table 3.5. The ORAMED project was conducted across 54 workers from six European countries and naturally the recorded data will be affected by differences in training and protection facilities in the different institutions. Indeed, it is acknowledged in the final ORAMED report that some ^{90}Y therapy procedures

Table 3.5. The range of maximum dose of all workers performing therapy procedures using ^{90}Y in the ORAMED project (Vanhavere *et al* 2012).

	Maximum normalised dose (mSv GBq^{-1})	
	^{90}Y-Zevalin	^{90}Y-Dotatoc
Preparation	1.2–43.9	0.1–7.4
Administration	1.0–11.9	0.4–4.9

were performed without shielding and pictorial examples of some observed bad practices are presented in the final report. In these cases very high doses can be expected given that the contact dose rate for a vial containing 1 GBq of ^{90}Y can be as high as 700 mSv min^{-1} (Delacroix *et al* 2002). Other workers have published similar values for maximum normalised hand dose during ^{90}Y therapy (Cremonesi *et al* 2006, Grassi *et al* 2008, Rimpler *et al* 2008, Grassi *et al* 2009).

The ORAMED project found that during dose preparation the maximum dose was located on the non-dominant hand for three-quarters of the procedures, with the tip of the index finger being the commonest site, (for 22% of all measurements). During administration the non-dominant hand was slightly more likely to receive the maximum dose (53% versus 47%), with the tip of the index finger again the commonest site (also for 22% of all measurements). The dose measured at the base of the index finger of the non-dominant hand showed correlation with the maximum dose, but the dose measured required multiplication by a factor of seven to estimate the maximum dose received by either hand. If the ring dosemeter is worn at the same position on the dominant hand the measured dose underestimated the maximum by a factor of ~22. The uncertainties in estimated doses using such factors will be large. Other studies report similar wide variations in dose distribution across the hands but did not find a significant difference in the average dose to each hand (Cremonesi *et al* 2006, Grassi *et al* 2008, Grassi *et al* 2009). Furthermore, Cremonesi *et al* (2006) reported estimated fingertip doses in the region of 12 Gy during an incident in which a vial containing 16.7 GBq of ^{90}Y was handled for only 10 s. The worker involved was wearing a ring TLD which recorded a dose of 70 mGy. These studies would suggest that ring TLDs are inappropriate for the assessment of maximum hand dose for this radionuclide because of the possibility of very high localised doses.

Given the large finger doses that could be incurred from handling high activity ^{90}Y sources (potentially exceeding the dose limit), the large variation in dose distribution, the uncertainty in published data relating to routine monitoring, and the unsuitability of ring dosemeters for measuring fingertip dose, the following regime is recommended.

Individual monitoring using finger stall TLDs on the tips of the thumb and first three fingers of each hand is recommended for all workers handling ^{90}Y therapy. The use of dosemeters measuring $H_p(0.07)$ is particularly important (Brasik *et al* 2007). The number of fingers monitored may be reduced once a pattern of exposure has been determined.

3.4.2.2 ^{177}Lu therapy

There are less published data for ^{177}Lu therapy than for ^{90}Y and it was not included in the ORAMED project. Available data suggest that finger doses are much lower than for ^{90}Y therapy (Bakker *et al* 2006, Sghedoni *et al* 2011). Typical fingertip doses of 50 μSv GBq^{-1} have been measured for pharmacists performing labelling procedures and 6 μSv GBq^{-1} for physicians involved in administering ^{177}Lu

treatments (Sghedoni *et al* 2011). This would suggest that for an annual workload of 100 patients, each being administered 7.4 GBq ^{177}Lu the expected average fingertip doses of a pharmacist and a physician would be ⩽ 40 mSv and ⩽ 5 mSv respectively.

The dose distribution across the hands appears to be less variable than it is for ^{90}Y, possibly because of the lower energy of the β-emissions (β_{max} = 0.49 MeV) and the additional γ-emissions at 113 and 210 keV (Delacroix *et al* 2002). Doses to the tips of the thumb, index and middle fingers of pharmacists performing labelling tasks showed little variation, although there appeared to be a slight increase in dose to the middle fingers of each hand of the physicians administering the treatments (Sghedoni *et al* 2011).

Bakker *et al* (2006) measured finger doses received during the labelling and preparation of ^{177}Lu-dotatate for 42 patient administrations. A maximum dose of 53 ± 12 μSv GBq^{-1} was recorded for the tip of the middle finger of the dominant hand. Doses were slightly lower for the non-dominant hand. It was estimated that for a worker preparing 100 administrations of 7.4 GBq the average annual dose to the tips of the fingers would be 23 ± 11 mSv, while the maximum dose could reach 40 mSv.

Although the expected dose to the fingers is much lower than for ^{90}Y therapy, some routine monitoring of finger doses should be considered to ensure techniques are optimised. A thin dosemeter calibrated for $H_p(0.07)$ should be used given the relatively low energy of the β-emissions from ^{177}Lu (the maximum depth of penetration of β-radiation in tissue is ~2 mm) (Brasik *et al* 2007, Gillet 2015). The optimum location for a routine monitor has not been determined. The dose distribution across the hands does not appear to be steeply contoured, but there is little published data on this issue although Bakker *et al* (2006) obtained a value of 1.65 for scaling the dose measured at the base to that measured at the tip of the finger.

> Finger stalls should be considered for staff preparing and administering ^{177}Lu who also administer ^{90}Y or other β-particle therapies. However, if ^{177}Lu is the main therapy being prepared and administered, ring dosemeters worn at the bases of the index fingers on both hands are likely to be sufficient for routine monitoring. Initial measurements with finger stall dosemeters are recommended to check dose levels and to determine an appropriate scaling factor to relate doses measured with ring dosemeters to the maximum values.

3.4.2.3 ^{223}Ra therapy

This radionuclide has only recently been introduced into clinical practice, so published data are scarce. However, approximately 95% of the emitted radiation energy is in the form of α-particles ⩽ 7.4 MeV, which will have a maximum range in soft tissue of 65 μm (Johnson and Birky 2011) and are therefore unable to penetrate the epidermis of the skin. Other associated emissions include low energy β-particles (β_{max}= 0.5 MeV) and some low abundance γ-radiation (0.14–0.84 MeV). The combination of low administered activities (typically ~4 MBq per patient) and relatively low abundance of radiation capable of penetrating the epidermis means the finger doses are likely to be relatively low. Gillet (2015) measured finger doses at the base and tip of the index fingers of each hand for a workload of 19 administrations over a three month period. The maximum recorded doses were 3.8 $\mu Sv\ MBq^{-1}$ and 2.5 $\mu Sv\ MBq^{-1}$ for the tip and base of the index finger respectively. These data suggest that for an annual workload of 100 patients (600 individual administrations of 4 MBq each) the finger dose from this procedure should be < 10 mSv.

> Given the relatively low finger doses involved with ^{223}Ra therapy, routine dose monitoring would not be required, unless the workload was excessively high. If monitoring is considered necessary, a finger stall dosemeter should be used, because of the short range of the emitted radiation.

3.4.2.4 ^{131}I therapy

Although ^{131}I has been in use for many years there are no definitive published studies focused on extremity dosimetry. To complicate matters further, there are a number of different therapy applications which involve ^{131}I, but which will result in it being handled very differently (in some cases no dispensing is involved) therefore making intercomparison of studies very difficult.

The contact dose rates with sources of ^{131}I can be relatively high (Delacroix *et al* 2002), so the need for finger dosimetry should be carefully considered at risk assessment or whenever procedures are reviewed. Wrzesień and Olszewski (2012) measured the dose at 19 locations across the hands of nine experienced workers unpacking, measuring and administering ^{131}I capsules at five different nuclear medicine departments in Poland. They found that for six of the workers there was no significant difference in measured doses between dominant and non-dominant hands, with the middle phalanx of the index finger of the non-dominant hand receiving the highest average dose of 27 ± 38 $\mu Sv\ GBq^{-1}$. Similar results were obtained by Morton and Greaves (2004) in the UK. For an assumed workload of three administrations per week of 500 MBq and one administration of 3.7 GBq per month the expected annual finger dose would be < 10 mSv in this instance. However, they also recorded dose factors of 2 and 7 $mSv\ GBq^{-1}$ for the remaining three workers (the tip of the middle finger and tip of the thumb of the right hand respectively). Using the same assumed workload as above would yield annual doses of 240 and 840 mSv respectively for these workers. Clearly, these are extreme cases and one would anticipate that such high dose readings from individual dosemeters would trigger a review of work practices and staff training prior to any dose limit being breached. However, the potential for high finger doses when working with ^{131}I capsules should not be overlooked. Other studies have returned similar factors for fingertip doses to staff handling ^{131}I in liquid form. Leide-Svegborn (2012) measured a value of 100 ± 70 $\mu Sv\ GBq^{-1}$ to the tip of the left thumb, while Pant *et al* (2006) recorded dose factors of 80 ± 14 $\mu Sv\ GBq^{-1}$.

The selection of an appropriate dosemeter for ^{131}I should be carefully considered as the skin dose from its β-emissions could be underestimated (Brasik *et al* 2007). The best location for a finger monitor has yet to be established and scaling factors for relating ring dosemeter values to maximum dose are not available, although factors in the range 5–7 have been suggested for ring dosemeters at the base of the index finger (Pant *et al* 2006), although no distinction is made between the dominant and non-dominant hands.

> Routine monitoring of finger dose for staff dispensing ^{131}I capsules is unlikely to be necessary (expected annual dose 5–10 mSv). However, initial assessments using ring (or stall) dosemeters are recommended to check that practices are appropriate. Whether this is for the dominant or non-dominant hand should be determined by observation. Where dose levels are found to be low, annual measurements are recommended to ensure safe working practices are being maintained.

3.4.3 Eye monitoring

Concerns over the proposed reduction in dose limit for the lens of the eye has prompted increased interest in eye dose monitoring in nuclear medicine departments. Results from $H_p(3)$ dosemeters worn on head bands have predicted annual doses of 1–2 mSv for typical iodine therapy workloads (Summers *et al* 2012). Given that the maximum depth of penetration of ^{131}I β-radiation in water is 2 mm (Johnson and Birky 2011), these measured doses will result from γ-radiation only. ^{177}Lu and ^{223}Ra also emit low energy β-radiation with a maximum depth of penetration in water < 3 mm and so their particulate emissions should not contribute to $H_p(3)$ measurements. At present there are no published data available for eye doses from either ^{177}Lu or ^{223}Ra therapy. There is a directional dependence of dose from high energy β-emitters such as ^{90}Y, so it is important to use a dosemeter calibrated in terms of $H_p(3)$ for measuring eye dose (Behrens *et al* 2017).

3.5 Use of electronic personal dosemeters in optimisation

Electronic personal dosemeters (EPDs) can be used to identify procedures that make significant contributions to staff doses where the work performed is variable. Download of data for later analysis can reveal periods when dose rate levels are high. Comparison of daily dose levels can identify tasks giving rise to higher dose levels, where staff rotate around different duties (Sylvain and Bok 2002, Bouzarjomehri and Tsapaki 2011). Such assessments are difficult with passive dosemeters worn for several weeks, where individual contributions cannot be separated. On shorter time-scales, the instantaneous reading will show up individual actions that make more significant contributions to a wearer's dose (Stewart 2015). However, for fuller analyses of dose data, records of activities performed will be required to allow cross-referencing the dose rate pattern with the tasks performed. This might be achieved in a radiopharmacy through recording sessions using a video camera (Whitby and Martin 2003). If periods of high dose rate can be identified, staff can be made aware of when they are more exposed and through education can improve their protection and reduce their overall dose.

Probes connected via a cable to a data logger are available that are small enough to attach to a finger (Whitby and Martin 2003, Martin 2016). Knowledge of the energy response is necessary if a representative measure of finger dose is required, and the importance of this will depend on the energy of the photon emissions being measured. The arrangement of the detectors and filters mean that the response of the probe will vary with angle, and Whitby and Martin (2003) describe assessments for one type of probe. However, for optimisation purposes, an approximate calibration may be sufficient, since it is identification of periods when dose rates are high relative to other times within the monitoring period that is important. Wearing of a rigid probe adjacent to the fingertip can be uncomfortable and trials to determine what is feasible should be undertaken. Whitby and Martin (2003) found that wearing a probe at the side of a finger was more acceptable for staff.

Figure 3.2 shows an example of results from sessions in a small radiopharmacy (a) where the syringe used for drawing up was not shielded and (b) where a syringe shield was used. It readily demonstrates that the reduction in exposure achieved through use of syringe shields outweighed any increase in exposure time. The use of electronic dosemeters in this way enables differences in operational techniques for single sessions and even single procedures to be assessed, for which differences would

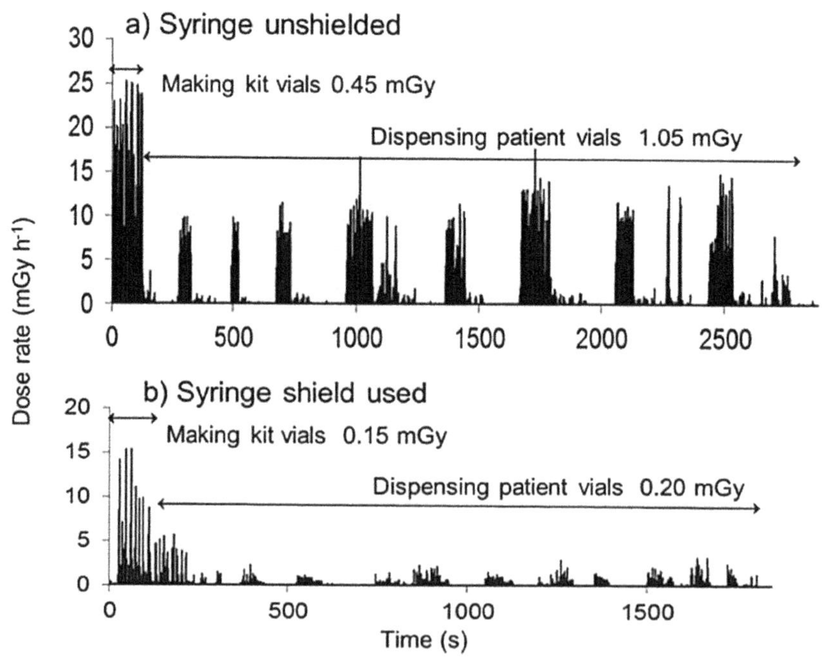

Figure 3.2. Variation in dose rate at the index finger of the dominant hand recorded by an extremity EPD probe during sessions dispensing methylene diphosphonate patient vials (a) with an unshielded syringe and (b) with a syringe shield.

be lost in results of monthly monitoring with passive dosemeters. Application of small probes in assessment of finger doses for nuclear medicine staff can demonstrate differences in dose resulting from injection technique, including finger positions, use of shielding, and prior insertion of a butterfly catheter into the vein that are difficult to appreciate from simple observations (Martin 2016). Electronic extremity dosemeters available currently have a limited energy and angular response that would prevent accurate measurement of maximum dose for radionuclides emitting β-particles.

References

Antic V, Ciraj-Bjelac O, Stankovic J, Arandjic D, Todorovic N and Lucic S 2014 Radiation exposure to nuclear medicine staff involved in PET/CT practice in Serbia *Radiat. Prot. Dosimetry* **162** 577–85

Bakker W H, Breeman W A, Kwekkeboom D J, De Jong L C and Krenning E P 2006 Practical aspects of peptide receptor radionuclide therapy with [177Lu] [DOTA0, Tyr3] octreotate *Q. J. Nucl. Med. Mol. Imaging* **50** 265–71

Batchelor S, Penfold A, Aric I and Huggins R 1991 Radiation dose to the hands in nuclear medicine *Nucl. Med. Commun.* **12** 439–44

Behrens R, Hupe O, Busch F, Denk J, Engelhardt K, Günther K, Hödlmoser H, Jordan M and Strohumaier J 2017 Intercomparison of eye lens dosemeters *Radiat. Prot. Dosimetry* **174** 6–12

Benatar N A, Cronin B F and O'Doherty M J 2000 Radiation dose rates from patients undergoing PET: impications for technologists and waiting areas *Eur. J. Nucl. Med.* **27** 583–9

Bilski P *et al* 2011 The new EYE-D™ dosemeter for measurements of $H_p(3)$ for medical staff *Radiat. Meas.* **46** 1239–42

Bouzarjomehri F and Tsapaki V 2011 Active personal dosimeter in a nuclear medicine centre in Yard City, Iran *J. Biomed. Phys. Eng.* **1** 28–33

Brasik N, Stadtmann H and Kindl P 2007 The right choice: extremity dosemeter for different radiation fields *Radiat. Prot. Dosimetry* **125** 331–4

Carnicer A *et al* 2011 Hand exposure in diagnostic nuclear medicine with 18F- and 99mTc-labelled radiopharmaceuticals—Results of the Oramed Project *Radiat. Meas.* **46** 1277–82

Chruscielewski W, Olszewski J, JankowskI J and Cygan M 2002 Hand exposure in nuclear medicine workers *Radiat. Prot. Dosimetry* **101** 229–32

Covens P, Berus D, Vanhavere F and Caveliers V 2010 The introduction of automated dispensing and injection during PET procedures: a step in the optimisation of extremity doses and whole-body doses of nuclear medicine staff *Radiat. Prot. Dosimetry* **140** 250–8

Cremonesi M, Ferrari M, Paganelli G, Rossi A, Chinol M, Bartolomei M, Prisco G and Tosi G 2006 Radiation protection in radionuclide therapies with ^{90}Y-conjugates: risk and safety *Eur. J. Nucl. Med. Mol. Imaging* **33** 1321–7

Dabin J, Kopeć R, Struelens L, Szumska A, Tomaszuk M and Vanhavere F 2016 Eye lens doses in nuclear medicine: a multicentric study in Belgium and Poland *Radiat. Prot. Dosimetry* **170** 297–301

Delacroix D, Guerre J P, Leblanc P and Hickman C 2002 Radionuclide and radiation protection data handbook 2nd edition *Radiat. Prot. Dosimetry* **98** 1–168

Demir M, Demir B, Yasar D, Sayman H, Halac M, Ahmed A, Ozcan K and Uslu I 2010 Radiation doses to technologists working with 18F-FDG in a PET center with high patient capacity *Nukleonika* **55** 107–12

Dhanse S, Martin C J, Hilditc T E and Elliott A T 2000 A study of doses to the hands during dispensing of radiopharmaceuticals *Nucl. Med. Commun.* **21** 511–9

Donadile L 2008 An Overview of the use of extremity dose meters in Some European Countries for medical applications *Radiat. Prot. Dosimetry* **1311** 62–6

Eleftheria C, Ginjaume M, O'Connor U, Kopeć R and Sans Merce M 2014 Status of eye lens radiation dose monitoring in European Hospitals *J. Radiol. Prot.* **34** 729–39

European Commission 2013 Council Directive 2013/59/Euratom of 5 December 2013 laying down basic safety standards for protection against the dangers arising from exposure to ionising radiation, and repealing Directives 89/618/Euratom, 90/641/Euratom, 96/29/Euratom, 97/43/Euratom and 2003/122/Euratom *Official J. Eur. Union* **57** L13 http://eur-lex.europa.eu/LexUriServ/LexUriServ.do?uri=OJ:L:2014:013:0001:0073:EN:PDF (assessed 26/4/16)

Gillet B 2015 *Finger Doses in Nuclear Medicine* Available at: http://www.ipem.ac.uk/Members/TrainingandOtherResources/IPEMConferencepresentations/PersonalDosimetry.aspx (accessed May 2017)

Grassi E, Sghedoni R, Asti M, Fioirini F, Salvo D and Borasi G 2009 Radiation protection in 90Y-labelled DOTA-D-Phe1-Tyr3-octreotide preparations *Nucl. Med. Commun.* **30** 176–82

Grassi E, Sghedoni R, Fioirini F, Asti M, Sarti M, Fraternali A, Versari A, Salvo D and Borasi G 2008 Extremity radiation protection of the staff in 90Y-DOTATOC therapy *J. Nucl. Med.* **49** 282

Guillet B, Quentin P, Waultier S, Bourrelly M, Pisano P and Mundler O 2005 Technologist radiation exposure in routine clinical practice with 18F-FDG PET *J. Nucl. Med. Technol.* **33** 175–9

Harding L K, Hesselwood S, Ghose S K and Thompson W H 1985 The value of syringe shields in a nuclear medicine department *Nucl. Med. Commun.* **6** 449–54

ICRP 2008 Working Party: Liniecki J, Jankowski J, Martin C and Mattsson S, Radiation exposure of hands in radiopharmacies: Monitoring of doses and optimisation of protection. ICRP Publication 106 *Ann. ICRP* 38 1–2, Annex E, pp 167-200 Elsevier

The Ionising Radiations Regulations 2017 *SI 2017 No 1075, London HMSO* http://www.legislation.gov.uk/uksi/2017/1075/contents/made (accessed 9/2/18)

Jodal L 2009 Beta emitters and radiation protection *Acta Oncol.* **48** 308–13

Johnson T E and Birky B K 2011 *Health Physics and Radiological Health* 4th edn (Baltimore, MD: Williams and Wilkins)

Kopeć R, Budzanowski M, Budzynska A, Czepczynski R, Dziuk M, Sowinski J and Wyszomirska A 2011 On the relationship between wholebody, extremity and eye lens doses for medical staff in the preparation and application of radiopharmaceuticals in nuclear medicine *Radiat. Meas.* **46** 1295–8

Leide-Svegborn S 2012 External radiation exposure of personnel in nuclear medicine from 18F, 99mTc and 131I with special reference to fingers, eyes and thyroid *Radiat. Prot. Dosimetry* **149** 196–206

Linemann H, Will E and Beuthien-Baumann B 2000 Investigations of radiation exposure of the medical personnel during F-18-FDG PET studies *Nuklearmedizin* **39** 77–81

Lundberg T M, Gray P J and Bartlett M L 2002 Measuring and minimizing the radiation dose to nuclear medicine technologists *J. Nucl. Med. Technol.* **30** 25–30

Magee J 2015 A study of occupational eye doses within nuclear medicine IPEM Personal Dosimetry Meeting (*Manchester*) http://www.ipem.ac.uk/Members/TrainingandOtherResources/IPEMConferencepresentations/PersonalDosimetry.aspx (accessed May 2017)

Martin C J 2016 Strategies for assessment of doses to the tips of the fingers in nuclear medicine *J. Radiol. Prot.* **36** 405–18

Martin C J and Whitby M 2003 Application of ALARP to extremity doses for hospital workers *J. Radiol. Prot.* **23** 405–21

Montgomery A, Anstee D E, Martin C J, Anstee D and Hilditch T 1999 Reductions in finger doses for radiopharmaceutical dispensing afforded by a syringe shield and an automatic dose dispenser *Nucl. Med. Commun.* **20** 189–94

Morton R J and Greaves C 2004 Finger and whole-body doses for a variety of therapy and diagnostic procedures *Nucl. Med. Commun.* **25** 41

Morton R J, Murray D, Zonoozi A, Pryor M, Plant G, Bellamy L and Hall J R W 2006 Are you measuring your true skin dose when handling FDG? [Abstract] *Nucl. Med. Commun.* **27** 1029

Pant G S, Sharma S K and Rath G K 2006 Finger doses for staff handling radiopharmaceuticals in nuclear medicine *J. Nucl. Med. Technol.* **34** 169–73

Peet D J, Morton R, Hussein M, Alsafi K and Spyrou N 2012 Radiation protection in fixed PET/CT facilities - design and operation *Br. J. Radiol.* **85** 643–6

Prince J and Murby B 2014 Eye dose measurements in a wide range of Nuclear Medicine departments indicate that it is unlikely that staff will need to be classified *Nucl. Med. Commun.* **34** 382

Rimpler A, Barth I, Baum R B, Senftleben S and Geworski L 2008 Beta radiation exposure of staff during and after therapies with 90Y-labelled substances *Radiat. Prot. Dosimetry* **131** 73–9

Roberts F O, Gunawardana D H, Pathmarai K, Wallace A, Paul L U, Tian M, Berlangieri S U, O'Keefe G J, Rowe C C and Scott A M 2005 Radiation dose to PET technologists and strategies to lower occupational exposure *J. Nucl. Med. Technol.* **33** 44–7

Seierstad T, Stranden E, Bjering K, Evensen M, Holt A, Michalsen H M and Wetteland O 2007 Doses to nuclear technicians in a dedicated PET/CT centre utilizing 18F fluorodeoxyglucose (FDG) *Radiat. Prot. Dosimetry* **123** 246–9

Sghedoni R, Grassi E, Fioirini F, Asti M, Piccagli V, Versari A and Iori M 2011 Personnel exposure in labelling and administration of ^{177}Lu-DOTA-D-Phe1-Tyr3-octreotide *Nucl. Med. Commun.* **32** 947–53

Smart R 2004 Task-specific monitoring of nuclear medicine technologists radiation exposure *Radiat. Prot. Dosimetry* **109** 201–9 *Reported at 11th International Congress of the International Radiation Protection Association* [online] available from http://irpa11.irpa.net/pdfs/5f40.pdf (accessed March 2016)

Stewart C 2015 Quantifying the eye doses of clinical staff working with ionising radiation and optimising protection methods *Personal Dosimetry IPEM Conference (21st January 2015)* (Manchester: Manchester Conference Centre and private communication) http://www.ipem.ac.uk/Portals/0/Documents/Conferences/2015/2015%20CONFERENCE%20ABSTRACTS.pdf (accessed May 2017)

Struelens L, Krim S and Vanhavere F 2010 *Extremity Doses of Medical Staff for Complex Interventional Procedures and in Nuclear Medicine ExDos, Final report 2010*. Available at: http://fanc.fgov.be/GED/00000000/2700/2757.pdf (accessed May 2017)

Summers E C, Brown J L E, Bownes P J and Anderson S E 2012 Eye doses to staff in a nuclear medicine department *Nucl. Med. Commun.* **33** 476–80

Sylvain I and Bok B 2002 Radiation exposure in nuclear medicine: real-time measurement *Braz. Arch. Biol. Technol.* **45** 111–4

Tandon P, Venkatesh M and Bhatt B C 2007 Extremity dosimetry for radiation workers handling unsealed radionuclides in nuclear medicine departments in India *Health Phys.* **92** 112–8

Vanhavere F *et al* 2012 ORAMED: Optimisation of radiation protection for medical staff *7th EURADOS Report* 2012-02 ISSN 2226-8057, ISBN 978-3-943701-01-2. Available at: http://www.eurados.org./en/Documents_Publications (accessed May 2017)

Walsh C, O'Connor U and O'Reilly G 2014 Eye dose monitoring of PET/CT workers *Br. J. Radiol.* **87** 20140373

Whitby M and Martin C J 2003 Investigation using an advanced extremity gamma instrumentation system of options for shielding the hand during the preparation and injection of radiopharmaceuticals *J. Radiol. Prot.* **23** 79–96

Whitby M and Martin C J 2005 A multi-centre study of dispensing methods and hand doses in UK hospital radiopharmacies *Nucl. Med. Commun.* **26** 49–60

Williams E D, Laird E E and Forster E 1987 Monitoring Radiation Dose to the Hands in Nuclear Medicine: the location of dosemeters *Nucl. Med. Commun.* **8** 499–503

Wrzesień M, Królicki L, Albiniak Ł and Olszewski J 2018 Is eye lens dosimetry needed in nuclear medicine? *J. Radiol. Prot.* **38** 763–74

Wrzesień M and Olszewski J 2012 Hand exposure of nuclear medicine workers during administration of radioiodine *Radiat. Prot. Dosimetry* **151** 147–52

Wrzesień M and Napolska K 2015 Investigation of radiation protection of medical staff performing medical diagnostic examinations by using PET/CT technique *J. Radiol. Prot.* **35** 197–207

Colin John Martin, David Temperton, Anthony Hughes and Thomas Jupp

Chapter 4

Dosimetry for personnel working with x-ray equipment

4.1 Background

Hospital staff wear protective lead/rubber aprons routinely when using x-rays. The lead aprons provide reasonable protection for organs within the trunk, while leaving the head, neck, arms and lower legs unshielded (Martin 2009). This non-uniform exposure presents unique challenges in obtaining assessments of personnel doses. Effective dose is usually estimated from the readings of personal dosemeters calibrated in terms of personal dose equivalent, $H_p(10)$. However, the reading from a single dosemeter placed above the lead apron will overestimate effective dose, while one worn under the apron will underestimate it. The overestimation by a dosemeter above the apron arises because it takes no account of the protection of organs in the trunk, while a dosemeter placed under the apron takes no account of higher doses to the head, neck, and organs in the thorax exposed via the apron's arm holes, leading to an underestimation (Franken 2002, Siiskonen *et al* 2007).

Following the latest revisions in the European Basic Safety Standard (EC 2013) and inclusion of the new eye dose limit in IRR17, staff working with x-rays are more likely to approach this limit rather than the one for the whole body, when it is protected by a lead apron. Therefore, any dosimetry system should include an assessment that can be related to the dose to the lens of the eye for staff working with x-ray equipment in controlled areas, who wear lead aprons, if doses levels are more than a few mSv.

This chapter contains a discussion of the problems involved in dose monitoring for staff using x-rays. Recommendations relating to monitoring requirements for different dose levels are included, while reviews of exposure and particular issues likely to be experienced with different areas of application are covered in chapter 5.

doi:10.1088/2053-2563/aaddb6ch4

4.2 Options for body dosimetry

In the UK, a single dosemeter has been worn on the body under the lead apron by the majority of staff using x-rays and the value of $H_p(10)$ recorded as this represents the dose received by the radiosensitive organs in the trunk. The reading of this type of dosemeter is the value entered into personal dose records for staff using x-rays as a measure of the effective dose. However, it should be remembered that this does not take account of doses to the head and neck, including the thyroid if a protective collar is not worn (Martin 2012, ICRP 2018), or the lungs and tissues exposed through the arm holes of a lead apron (Franken 2002). The dosimetry method recommended by the International Commission on Radiological Protection (ICRP) for more highly exposed interventional radiology and cardiology staff is for two dosemeters to be used (ICRP 2000, 2013, 2018), 'since interventionalists typically receive the highest occupational doses, and the actual attenuation of the interventionalist's apron is rarely known'. One is worn under the lead apron to assess doses to organs in the trunk, which make up the majority of tissues known to be more radiosensitive, and one is worn at the collar to record a measure of doses to the head and neck. However, there is no international consensus on the approach adopted for dose monitoring of x-ray users. Different European countries recommend use of a single dosemeter worn under the apron, a single dosemeter worn at the collar above the apron, or two dosemeters one above and one below the apron (Järvinen 2008a). The US National Council on Radiation Protection and Measurement (NCRP) recommend the two dosemeter option for interventional staff, but state that a single dosemeter worn at the neck above protective garments may be used (NCRP 2011). The ICRP state that 'if, for a given institution, it can be reliably assumed that all professionals wear protective aprons with sufficient attenuation during all interventions, the under-apron dosemeter could be omitted for the majority of staff members of the interventional team, because the over-apron dosemeter will be sufficient for the assessment of their exposure' (ICRP 2018). Wearing of a single dosemeter under the apron is not recommended, as it will not provide any information relating to the dose to the eyes, and for staff who do not work close to the x-ray source the majority of dosemeter readings recorded will be below the minimum detection level. Therefore the accuracy of the technique is poor and the value in providing radiation protection information is limited. However, for staff who perform some of their duties while wearing a lead apron, but do not wear one all the time, then it may be important to make a measurement using a dosemeter worn on the body to assess effective dose, and ensure that this dosemeter is covered by the lead apron when one was worn.

4.3 Assessment of eye dose

For assessing the equivalent dose to the lens of the eye, the operational quantity personal dose equivalent at a depth of 3 mm ($H_p(3)$) should be used where possible, as recommended by the International Commission on Radiation Units and Measurement (ICRU) (Behrens and Dietze 2011, Behrens 2012). A number of different dosemeters are now available within Europe (Carinou *et al* 2015). The majority use a single detector either attached to a headband or with a clip to allow it to be attached to protective eyewear (Cantone *et al* 2017). Dosemeters should ideally be positioned on the head near to the eye. The superficial dose $H_p(3)$ to a dosemeter worn at the collar above the lead apron can be used as an indicator of eye dose, as part of the risk assessment process and can be used for routine monitoring if this is an approved eye dosemeter. Where a risk assessment indicates a possible need for eye monitoring, staff may be monitored with a dedicated eye dosemeter for a trial period and comparisons made with results from a collar dosemeter in order to evaluate options for future monitoring. If the dosemeter is only calibrated for $H_p(0.07)$, this can be used as an alternative for non-classified workers, as it will give a reasonable indication of eye dose with x-ray exposures.

Those more likely to approach the eye dose limit are interventional radiologists or cardiologists. Training in effective use of ceiling suspended screens can aid in reducing eye dose levels, and improvements will be reflected in dosimetry results, so eye dose monitoring should be able to assist in demonstrating good practice. However, these groups may also wear lead glasses, but the protection offered will not be apparent unless the dosemeter is also shielded. The wearing of dosemeters behind lead glasses is difficult, and they do not necessarily give a good indication of the dose to the eye lens. If the dosemeter is worn outside the lead glasses, the protection provided should be taken into account in assessment of the dose to the eye lens and a protection factor could be applied to take account of the dose reduction. The protection afforded by lead glasses depends on the angle at which scatter from the patient is incident on the head. For the majority of the time that an interventional clinician is carrying out a procedure, he/she will not be looking towards the patient when x-rays are being emitted. Therefore the protection factor required must take account of x-ray beams incident from the side and below the level of the head. Dose reduction factors in clinical practice are usually about four or five (McVey *et al* 2013, Van Rooijen *et al* 2014, Magee *et al* 2014). The dose to the eyes when protective eyewear is worn results primarily from radiation scattered from surrounding tissues (Cousin *et al* 1987, McVey *et al* 2013, Magee *et al* 2014, Ferrari *et al* 2016). The dose reduction factor is proportional to the size of glass lenses for areas between 6 and 30 cm^2 if lead equivalence is the same (Hu *et al* 2017). The closeness of the fit of the glasses and the size of the lenses protecting regions beneath the eye determine the protection factor. The application of a protection factor of 0.5 (dose reduction factor = 2) when lead glasses of standard design are used is a reasonable, conservative assumption (Magee *et al* 2014, ISO 2015). Whether or not this method can be used for the official dose record will depend on whether ADSs obtain approval from HSE for a dosemeter with the appropriate designation.

4.4 Use of a collar dosemeter worn outside lead protection

4.4.1 Assessment of body dose

4.4.1.1 Staff who always wear a lead apron when exposed to x-rays

When all doses are low, what is required is a dosimetry system that provides a reasonable indication of dose levels and is simple to implement. A single dosemeter worn outside the collar of the lead apron gives a measure of the radiation level in the environment (figure 4.1(a)). It can provide an indication of eye dose, and some information on body dose for staff who always wear lead aprons when working with x-rays. Such a dosemeter gives an overall indication of radiation levels to which staff involved in radiology procedures are exposed. Account can then be taken of the likely reduction in dose for parts of the body protected by the lead apron. Thus this approach is generally suited to monitoring of non-classified staff working with x-ray equipment who always wear lead aprons when working with x-rays and practices in the UK should ensure that exposure when not wearing any protection is minimal.

The $H_p(10)$ reading for a collar dosemeter worn with a lead apron will give the exposure of organs in the head and neck, and when no thyroid collar is worn, there is a high degree of correlation ($P > 0.9$) between a measurement of the entrance dose at the neck and the effective dose, as the doses to the thyroid and head make a major contribution (Kicken *et al* 1999). The correlation between effective dose and the unshielded dosemeter measurement is even stronger for the more recent ICRP definition of effective dose (ICRP 2007), because other tissues in the head have higher weighting factors (von Boetticher *et al* 2008, 2009).

Studies of the relationship between the dose at the neck and values for effective dose derived from simulations suggest that application of a correction factor between 0.011 and 0.18 to the $H_p(10)$ measurement would give a reasonable assessment of effective dose for situations where an apron is worn without a thyroid collar, and factors between 0.02 and 0.083 when both an apron and a thyroid collar are worn (Martin and Magee 2013). Therefore, multiplication of the collar dose by a factor of 0.1, which is easy to remember and simple to apply, can provide a conservative approach for estimating the likely effective dose for the majority of staff working with x-rays (ICRP 2018). This factor can be used for aprons with lead equivalences of 0.25 mm or 0.35 mm, as differences are small when compared to the overall uncertainty. This application of a factor of 0.1 can be employed as an

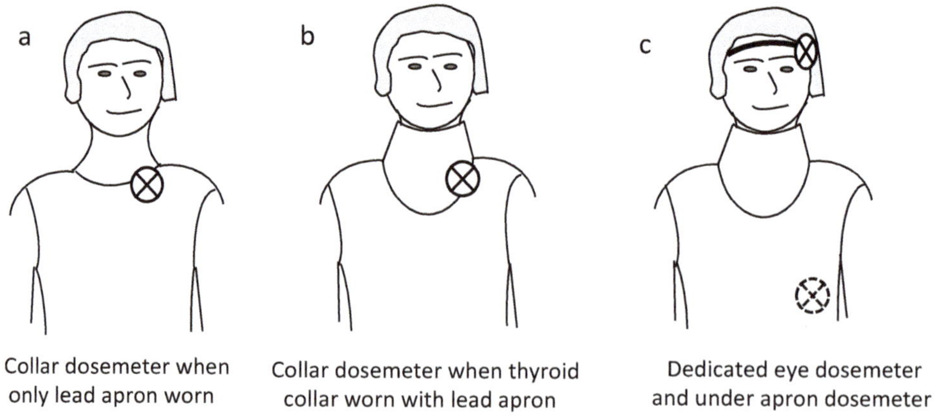

Figure 4.1. Positions of dosemeters for monitoring of staff exposed during x-ray procedures. The cross inside a circle shows the position of the dosemeter and when this is drawn with dashed lines, it indicates that the dosemeter is worn under the lead apron.

indicator by the local radiation protection service when reviewing dosemeter readings on behalf of the employer, but because it is an estimate, it is not suitable for recording in the official dose record for classified workers.

4.4.1.2 Staff who may be exposed to x-rays with or without a lead apron being worn

Some staff may be exposed in different situations wearing a lead protective apron on most, but not all, occasions when there is a risk of radiation exposure. If this is considered to be the case, then a dosemeter worn on the body, which is positioned under the lead apron when one is worn, is likely to be the preferred option to assess effective dose. The appropriate option for monitoring of individual staff should be determined in the risk assessment. However, it is recommended that for any department a standard practice should be adopted for all staff members.

4.4.2 Assessment of eye dose

For assessing the equivalent dose to the lens of the eye, the operational quantity personal dose equivalent ($H_p(3)$) recorded by the collar dosemeter should be used. A correlation has been found between collar dose and eye dose for interventional clinicians (Martin and Magee 2013, Farah *et al* 2013, Principi *et al* 2015, Higgins 2016, Fetterly *et al* 2017), but results can be variable because the radiation field is non-uniform, and interventional operators use ceiling suspended screens which may shield the eyes but not the neck. However, for the majority of staff, who will be further from the x-ray source, the radiation field will be more uniform, so the collar dose (figures 4.1(a) and (b)) is likely to be a reasonable reflection of the eye dose. Thus a collar dose is a useful indicator for whether a dedicated eye dosemeter (figure 4.1(c)) should be worn, and a collar dosemeter is a suitable first dosemeter for the majority of radiology staff. Ratios of eye and collar doses for interventional operators, for whom the exposure is going to be more non-uniform gave median and third quartile values of 0.6 and 0.9 respectively (Martin and Magee 2013, Higgins 2016), phantom studies gave values between 0.59 and 0.75 (Fetterly *et al* 2017), and Monte Carlo simulations have given a ratio of about 0.75 (Clerinx *et al* 2008). Thus a collar dosemeter can give an indication of the exposure of the eyes as well as the body. It would be appropriate for all staff working with x-rays to treat the $H_p(3)$ reading from a collar dosemeter as a measurement of eye dose. If the collar dosemeter reading exceeds 10 mSv per year, then the controls in place to ensure that doses are as low as reasonably practicable (ALARP) should be reviewed and wearing of a dedicated eye dosemeter is recommended.

4.5 The double dosemeter body monitoring system

Wearing of two dosemeters one either adjacent to the eye or at the collar above the lead apron, and one underneath the apron provides the best option for assessment of effective dose for more highly exposed workers, and is the method recommended by the ICRP for monitoring of interventional staff (ICRP 2000, 2013). However, where all staff present in the interventional suit can reliably be assumed to always wear their lead aprons, ICRP acknowledge that two dosemeters are only necessary for the

Table 4.1. Coefficients for use in equation (4.1) to calculate effective dose.

Algorithm	With thyroid shielding		Without thyroid shielding	
	α	β	α	β
Swiss Ordinance (2008)	1	0.05	1	0.1
Von Boetticher *et al* (2010)	0.79	0.051	0.84	0.100
Recommended	1	0.05	1	0.1

interventionalist performing the procedure; and it is sufficient for other staff to wear an over-apron dosemeter at collar level (ICRP 2018).

A number of different algorithms have been developed to combine readings from a collar and an under-apron dosemeter over the years to provide a better assessment of effective dose. Although this approach is unlikely to be adopted in the UK, some discussion is included here, as the method is widely used in many parts of the world. A variety of algorithms have been developed to evaluate effective dose (E) from the readings of two dosemeters and the form can be represented by the expression:

$$E = aH_{ub} + bH_{on} \tag{4.1}$$

where H_{ub} and H_{on} are the personal dose equivalents $H_p(10)$ measured on the body under the apron and at the neck outside the apron, respectively. H_{ub} is measured either on the chest or the waist, while H_{on} is generally measured at the collar. H_{on} could also be taken from a dosemeter worn for an eye dose assessment. Several studies have been carried out comparing different algorithms, but there has been no agreement on which provides the best option. The study by Järvinen (2008b) concluded that there is no algorithm which is optimum for all possible geometries and that therefore, compromises have to be made when making a choice. From the double dosemeter algorithms tested, the one recommended in the Swiss Ordinance (2008) is as close as any to giving a realistic value for effective dose. More recently algorithms based on ICRP (2007) tissue weighting factors have been developed using $H_p(10)$ at the neck measured outside the apron and at the trunk under the apron (von Boetticher *et al* 2010). Values from both of these studies are presented in table 4.1. The factors employed in the Swiss Ordinance are recommended, as they are simpler to remember and apply, and are the more conservative values.

The accuracy that can be achieved relies on interventional staff wearing the two dosemeters correctly and consistently, and ensuring that the positions of the dosemeters are never reversed. In the UK, any such algorithm would only be used to provide the local radiation protection service with a better measure of effective dose, so the value recorded by the dosimetry service for effective dose will be that of the under-apron dosemeter.

4.6 Recommendations for dose monitoring

4.6.1 Non-classified workers

Personal dose monitoring requirements should be determined through a risk assessments based on each individual's work. Risk assessments are discussed in more detail in chapter 7, but guidance on likely requirements for personal dosimetry based on dose levels in x-ray applications is given in table 4.2.

> Monitoring of the dose to the eye lens for staff working with x-rays can be achieved with either a dedicated eye dosemeter or a dosemeter worn at the collar outside the lead apron. If the staff member is involved directly in x-ray procedures, the dosemeter should be worn on the side adjacent to the x-ray tube for the majority of these procedures (figure 4.1).

If the staff member always wears a lead apron, the collar dosemeter can also be used to demonstrate whether the wearer is likely to require a body dosemeter under their apron and a dedicated eye dosemeter as set out in the flow chart in figure 4.2.

The employer's radiation protection service can use the following relationships to obtain approximate assessments of effective dose (E) and eye lens dose (D) from personal monitoring results for a collar dosemeter (Martin and Magee 2013).

$$E = 0.1 \times H_{on} \tag{4.2}$$

$$D = H_p(3)\left(\mathrm{or} H_\mathrm{p}(0.07)\right) \tag{4.3}$$

Assessments carried out with a collar dosemeter (i.e. for non-classified staff) using equation (4.2) as an indicator of effective dose are supported by the ICRP (ICRP 2018), but because the measurement is an approximation, the result is not currently formally approved by HSE as a measurement of effective dose. As such any dose assessments entered into the dose record kept by a coordination and record keeping service would need to be entered as estimates only. However, as long as this is made clear to the employer, this will be satisfactory as there is no legal requirement for an ADS to keep records for non-classified staff.

> Staff working with x-rays who record more than 10 mSv on a collar or eye dosemeter are recommended to use a dedicated eye dosemeter and a dosemeter under their lead apron (figure 4.1(c)).

The majority of staff working with x-rays are likely to wear a lead protective apron whenever they are working in a controlled area directly with x-ray equipment. At other times, for example when they are operating radiographic equipment, they will be within a protected operator cubicle. Therefore, a collar dosemeter should provide an acceptable record of their doses.

However, there may be certain staff who work between Radiology and another radiation department such as Nuclear Medicine, where they will not wear a lead apron routinely, but will still be exposed to radiation. A body dosemeter would probably be required for Nuclear Medicine, and in this case it would be appropriate

Table 4.2. Proposed dose levels for implementation of dose monitoring for staff working with x-rays.

Tissues	Dosemeter position	Dose quantity normally reported	Annual dosemeter reading (mSv)	Monthly dosemeter readings (mSv)	Dose monitoring recommendations
Eyes	Collar or headband	$H_p(3)$	1–5[a]	0.1–0.4	Initial monitoring with collar or head dosemeter to establish dose levels.
Eyes	Collar or headband	$H_p(3)$	6–10[a]	0.5–1	Regular monitoring with collar or head dosemeter recommended.
Eyes	Headband	$H_p(3)$	⩾10[a]	⩾1	Regular monitoring with head dosemeter recommended and classification likely.
Body	Option 1 Collar above apron	$H_p(10)$	⩾6	⩾1	Regular monitoring with additional under-apron dosemeter is recommended.
Body (mixed work)	Option 2 Under-apron	$H_p(10)$	⩾1	⩾0.2	Regular monitoring with under-apron dosemeter is recommended.
Hands	Finger	$H_p(0.07)$	20–50	2–4	Assessment of most exposed area and initial monitoring to establish dose levels.
Hands	Finger	$H_p(0.07)$	50–100	4–10	Regular monitoring with finger dosemeters should be considered. Establish ratios to correct dose, if ring dosemeter does not measure most exposed area. (see section 4.6.5).
Hands	Finger or wrist	$H_p(0.07)$	⩾100[b]	⩾10	Regular monitoring with finger dosemeters is recommended.
Legs	Lower leg or feet	$H_p(0.07)$	>100	>10	Initial monitoring to establish dose levels and assess protection options.

[a] The recorded dose divided by 2 may be used as the indicator, if protective eyewear is worn consistently by the staff member.
[b] After application of any required factor to correct for position of the dosemeter.

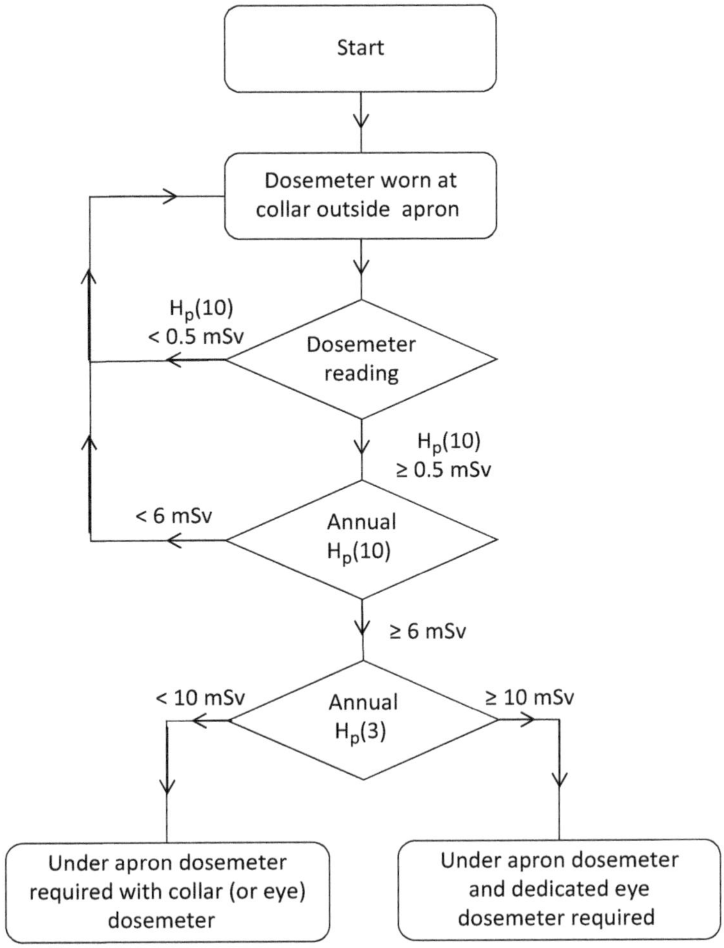

Figure 4.2. Flow chart to illustrate when different dosemeters should be worn by staff working with x-rays and wearing lead aprons.

for it to be worn under the lead apron when the staff member was working with x-rays, and the $H_p(10)$ value recorded as the effective dose. A second dosemeter may be required to measure eye lens dose in this case, if the x-ray exposure was likely to be significant.

> Since confusion will arise if more than one dose monitoring system is in place, for staff who wear a single dosemeter, it is recommended that a similar arrangement is adopted for all staff within a department who work with x-ray equipment.

The appropriate options for monitoring staff should be determined through risk assessments. Risks of doses to the eyes and of effective doses to staff exceeding a few mSv should be considered and the system that is more appropriate to the department or organisation adopted.

> If a change to use of a collar dosemeter is made, it is important that this is done in collaboration with the ADS and that all staff members are made aware of the implications for dose recording. Matters that should be considered when making such a change are discussed in section 4.7.

4.6.2 Classified workers

> Classified staff working with x-rays are recommended to use two dosemeters, one preferably on the head adjacent to the eye or alternatively at the collar of the lead apron measuring $H_p(3)$, and the second under the lead apron recording $H_p(10)$. All these dosemeters must be provided as part of an approved service.

The under-apron $H_p(10)$ measurement will be recorded as the effective dose in the dose record and the eye or collar dosemeter $H_p(3)$ measurement recorded as the eye dose. The ADS will report its doses to an approved coordination and record keeping service. It is this dose that forms the statutory dose record. For any system used the ADS must have approval from the HSE, and the method must be included in the employer's statement of service.

When two dosemeters are worn, it is important that each is always located in the correct position. If the two dosemeters are different types, e.g. a body dosemeter under the apron and a dedicated head dosemeter, then the two should not be confused. However, if a collar dosemeter of similar type to the under-apron dosemeter is used, there is a risk that staff may put the under-apron one in the wrong position. Even if this is only for one case, it can have a significant impact on the assessment effective dose, since the exposure received by the unshielded dosemeter may be ten times that of the under-apron one. The better option is to use a dedicated eye dosemeter, but if a collar dosemeter is used, methods to allow ready identification of each dosemeter such as colour coding should be employed.

4.6.3 Body dosemeter design

All methods of attachment should be strong enough to prevent the dosemeter being dislodged during strenuous use. The comfort in wearing should be an important element when choosing suitable dosemeters, as any annoyance is likely to affect compliance in wearing them. For whole body dosemeters placed near the neck at the top of the lead apron (figure 4.1(a)) or over a protective thyroid shield, consideration should be given to identifying any edges of the dosemeter that could irritate the neck or chin area and action taken to minimise discomfort. If a thyroid collar is worn, then the dosemeter might be attached to the lower edge of the collar (figure 4.1(b)). Dosemeters worn under lead aprons should not induce sharp pressure points that may cause discomfort when placed between the heavy leaded apron and the user's clothing.

4.6.4 Eye dose monitoring

4.6.4.1 Eye dose measurement

When a risk assessment indicates that the annual dose to the eye may be in the range 6–10 mSv, then a method for obtaining a regular assessment of eye dose is required. The $H_p(3)$ dose recorded by a collar dosemeter may be used for this purpose, but use of a dedicated eye dosemeter should be considered.

> If the annual dose measured by a collar dosemeter exceeds 10 mSv, staff are recommended to wear a dedicated eye dosemeter measuring the dose to the more exposed eye (figure 4.2 and table 4.2).

Monitoring of the lens of the eye presents special challenges due to the absence of objects near the eyes on which to attach the dosemeter. Dosemeters should ideally be positioned on the head near to the eye. Factors relating to technique, geometry, x-ray beam quality and use of protective devices all affect doses to the eyes of interventional clinicians. Dosemeters worn on the head at the side of the eye or on the eyebrow ridge adjacent to the x-ray tube (usually the left) or in the middle of the forehead can all provide reasonable assessments of eye doses. Many studies have shown that the dose at the side of the eye is higher than that at the forehead (Carinou *et al* 2015), but this is not always the case and the optimum dosemeter position may vary with the type of procedure and the operator (Vanhavere *et al* 2012, Domienik *et al* 2014, Efstathopoulos *et al* 2011, McVey *et al* 2013). The acceptability of wearing head dosemeters and ensuring that interventional clinicians wear them consistently are significant issues for radiation protection. Employers will need to have systems in place to ensure compliance, including personal monitoring policies, clear instructions in local rules, periodic audits of dosemeter use, and disciplinary processes to deal with non-compliance.

Dedicated eye dosemeters will generally be worn on a headband (Bilski *et al* 2011, PHE 2014) and this must be tight enough to keep the dosemeter in the proper position, and should not obscure vision. One option that may help in some circumstances is to attach the dosemeter to lead glasses, or even incorporate it within the side arm of the glasses, but this has the disadvantage that it requires the lead glasses to be worn whenever the individual is exposed to radiation. If this improves compliance in wearing dosemeters this may be a useful option, but caution should be exercised in interpretation of results, as explained in section 4.6.4.2. The need to maintain the sterility of the dosemeter and the holder that are used on a daily basis is also a concern.

4.6.4.2 Dosimetry when wearing protective eyewear

If operators wear eye protection, then this should be taken into account in assessment of the dose to the eye lens. The dosemeters available that can be worn behind lead glasses are limited at the present time, but ones that allow the dosemeter to be positioned on the skin surface at the side of the eye may provide a good option (IRSN 2018). Dosemeters incorporated into lead glasses immediately behind the protection or clipped onto the inside of the lenses of lead glasses may be shielded from x-rays coming from the side that will be incident on the eye, and so may significantly underestimate the dose to the eye lens (Martin 2018). A dosemeter attached to the frame of the lead glasses outside the protection provides a possible option, but this will overestimate the dose from incident radiation, while not recording all the radiation scattered back from tissues in the head, so the net result may be closer to the eye lens dose but the uncertainties will be large (Silva *et al* 2017). The dose measured with a dosemeter attached to a headband is likely to give a more representative result, but will overestimate the eye lens dose.

If a dosemeter is worn on the skin surface near to the eye, but not shielded by the lead glasses, a protection factor could potentially be applied to take account of the dose reduction. The size of the lenses protecting regions beneath the eye and the closenesss of the fit of the glasses determine the protection factor (see section 4.3). A factor of 0.5 would be a reasonable conservative value for the majority of users (Magee *et al* 2014, ISO 2015, ICRP 2018).

> If practicable, dosemeters that can be worn on the skin surface behind protective eyewear should be used. If suitable dosemeters are not available, the application of a protection factor of 0.5 to the measurement of eye dose could be used for staff who wear lead glasses of standard design.
>
> For any protection factor to be applied, the employer must have a documented quality assurance programme in place to confirm that the staff member concerned wears both the protective eyewear and the dosemeter in the designated position for all procedures. Adjustment or correction factors cannot be applied to the dose records by an ADS until the process has been included in the ADS's statement of service and approved by HSE.

If a measurement of the protection factor in clinically realistic configurations can be made, then a lower protection factor between 0.3 and 0.4 might be used (Magee *et al* 2014). For non-classified wearers such protection factors could be applied by the employer to results for eye dose or where a collar dose is used as a surrogate for eye dose. As before, the dose entered into a classified wearer's dose record by the approved coordination and record keeping service forms a statutory dose record. If correction factors for personal protective equipment (PPE) are applied to the assessed doses for classified wearers they cannot be applied to the dose records by an ADS until the process has been included in the ADS's statement of service and approved by HSE. The employer must have a system in place to confirm that both the lead glasses and the dosemeter are always worn in the designated position.

4.6.5 Hand dose monitoring

Interventional clinicians, orthopaedic surgeons and other healthcare staff working with mobile C-arm equipment use x-ray imaging to aid the procedures, so their hands need to be close to the x-ray field. The dose rate within the primary beam transmitted through the patient would typically be 2–5 μGy s^{-1}, so if their hand strays into the imaging field, a one minute exposure would give a dose of 100–300 μGy. Dose rates from primary beam x-rays scattered directly from the surface of the patient will be much higher and direct exposure to the incident primary beam in an overcouch configuration could be 50–100 times greater. Thus doses for these groups of staff can readily exceed 100 mSv. In configurations with the x-ray tube above the patient, even short exposures to the primary beam can give significant doses and must be avoided (Domienik *et al* 2014).

The dose limit for the skin is averaged over an area of 1 cm^2, so it applies to the most exposed part of the hand and it is important to position the dosemeter correctly. Small ring shaped dosemeters usually provide the best option for interventional radiologists and cardiologists. Ring dosemeters are available in several sizes for different finger diameters. Expandable plastic rings have been known to become tight on larger fingers that may swell during long procedures, so too tight a fit should be avoided.

In procedures performed by some groups such as orthopaedic surgeons, it may be the tip of the finger that receives the highest dose (Hafez *et al* 2005), and in this case a finger stall or sheath dosemeter which positions the detector element at the tip should be considered, but this may be unacceptable to staff because of the reduction in tactile sensation that results. Placing a ring dosemeter on the second phalanx is a possible option in this case. In most x-ray procedures the hand is exposed to scatter from a large x-ray field, so the dose to the fingertip should not be substantially greater than that to the base of the finger, unless the operator puts their hand into the primary beam. There are specific issues relating to interventional CT and these are discussed in section 5.3.3.2.

Infection control is an essential requirement for most procedures for which a finger dosemeter would be worn. Concerns about whether dosemeters can withstand a sterilisation process, often leads to their being used under surgical gloves. But a sterile glove worn over the dosemeter may not be sufficient to ensure adequate infection control, as the glove may develop holes during a procedure. Kopeć *et al* (2016) report favourable results for chemical sterilisation of waterproof extremity dosemeters, but some changes were observed for ones where the element was uncovered that depended on the chemical used. Decreases in response of ~30% were reported following steam sterilisation of TLDs. Therefore care is required in deciding upon the most appropriate practice to adopt and this may involve discussions with infection control staff as well as the personnel being monitored.

4.7 The change to collar dosemeters

The change to a dosimetry system based on a collar dosemeter worn outside any protection as the primary measurement for non-classified staff working with x-rays would be a major shift from current practice. It would involve an increase in the magnitude of doses that staff record so if the risk assessment indicates that such a change should be made, the consequences of, and reasons for, the change need to be fully explained prior to the change being introduced.

The primary reason for the change is to provide a measurement related to the dose to the lens of the eye. But a collar dosemeter will give a measurement of the radiation level to which a person is exposed, so it can be used to give an approximate indication of the effective dose by application of the adjustment factor. However, as discussed earlier, a complicating factor is that some staff may only wear personal protective equipment for part of the time. For example the majority of the work for general radiographers will be within protected cubicles, but they may carry out some work with mobile radiography or fluoroscopy that will require them to wear lead coats. The doses received when staff are not wearing lead aprons should be small, provided proper protection measures are being employed and staff are trained in the use of both the protective equipment and the dosemeters, therefore wearing a lead apron and following the system set out in the flow chart in figure 4.2 will be appropriate.

Practices among different groups will vary and typical examples are given in table 4.3. The collar dosemeter result will give a substantial overestimate of the effective

Table 4.3. Practices in wearing of lead aprons for different groups of hospital staff using x-rays and suggested approaches to treating results from collar dosemeters in relation to the limit on effective dose (E) in the period after the change is made.

<table>
<tr><th>Staff group</th><th>Practice in wearing lead aprons</th><th>Approach to monthly $H_p(10)$ dosemeter measurement</th></tr>
<tr><td>General radiographers</td><td>Mostly in protected cubicle, but some work wearing lead aprons</td><td rowspan="4">If $H_p(10) \geqslant 0.5$ mSv, initial investigation to check whether staff have been wearing lead aprons whilst working in x-ray theatre or attending CT patients during scans. If they have and will continue this work on a regular basis, no further action is required unless monthly reading > 1 mSv or summed annual dose approaches 6 mSv. Then use of separate dosemeters under the apron and adjacent to the eye should be considered.</td></tr>
<tr><td>Radiologists</td><td>Wear lead aprons occasionally</td></tr>
<tr><td>X-ray nurses</td><td>Mixed pattern of work including interventional and theatre for which lead aprons are worn</td></tr>
<tr><td>CT radiographers</td><td>Mostly in protected cubicle, but wear lead aprons occasionally</td></tr>
<tr><td>Interventional radiology staff</td><td rowspan="3">Wear lead aprons for all x-ray work</td><td rowspan="3">$E \approx 0.1 \times H_p(10)$ investigate if $H_p(10) \geqslant$ 1 mSv or summed annual dose approaches 6 mSv. Then use of separate dosemeters under the apron and adjacent to the eye should be considered.</td></tr>
<tr><td>Clinicians using mobile C-arms in theatres</td></tr>
<tr><td>Vascular theatre staff</td></tr>
</table>

dose when a lead apron is worn. For the groups, to whom this change is applied, that wear lead aprons some of the time, this will initially require checks to be made of their work patterns during the period when the dosemeter was worn, if the monthly $H_p(10)$ result exceeds 0.5 mSv. One option might be for staff members for whom this might apply to wear collar and under-apron dosemeters for a trial period.

Some departments have already made a change to a collar dosemeter, so it is possible to gain some insight into dose levels from the initial results. McCallum *et al* (2015) report a study in which collar dosemeters were issued to a variety of staff, coupled with setting of investigation levels to trigger follow up, and found that after the changeover the vast majority of staff still recorded doses below 0.1 mSv per wear period. The main exceptions were nursing staff working a rotation including x-ray theatres, of this group about 6–8 nurses per month out of 40 were over 0.3 mSv which had been set as an initial investigation limit. The nurse manager confirmed in each case that the nurses were working either in cardiology catheterisation laboratories or x-ray theatres during this time and therefore wearing lead coats throughout periods of exposure to x-rays, so no further investigation was necessary. Once dosimetry patterns have been established, realistic investigation levels should be set for each group of workers.

If proper preparations are made prior to introduction of the collar dosemeter, then it should be possible to manage the transition and avoid any major issues. The change provides an opportunity to educate staff about the effects of radiation on the eye at the same time as highlighting the need to alter and improve dosimetry practices. Moreover, the fact that measurable dose results are recorded should assure staff of the value of wearing personal dosemeters, as well as providing radiation protection practitioners with valuable information about dose levels in different environments within the hospital.

Table 4.4. Investigation levels and action levels based on dosemeter readings (before any corrections applied).

Dosemeter position	Dose quantity	Annual dose investigation levels (mSv)		Monthly dose action levels (mSv)	
(and eye protection)		Other x-ray users	Interventional clinicians	Other x-ray users	Interventional clinicians
Eye or collar with no eye protection	$H_p(3)$	6	10	1	2
Eye dosemeter behind lead glasses	$H_p(3)$	6	10	1	2
Eye or collar, eye protection worn with dosemeter outside[a]	$H_p(3)$	10	15	1	2
Body over-apron[b]	$H_p(10)$	6	15	1	2
Body under-apron	$H_p(10)$	1	2	0.5	1
Hands	$H_p(0.07)$	—	100	—	20
Legs and feet	$H_p(0.07)$	—	100	—	20

[a] Agreed protection factor applied subsequently.
[b] Correction factors in equation (4.2) would be applied subsequently where this is agreed.

4.8 Investigation levels

Investigation levels provide a tool to ensure that individual doses are highlighted, so that action can be taken to ensure that doses are maintained ALARP. They are generally set in terms of the annual dose received during a calendar year. Values need to be established against which managers can make comparisons and decide on whether further investigation is required. The regulations require that whole body investigation levels are set in the local rules. Suggested values that might be used are given in table 4.4, but these should be adjusted according to local practices and dosemeter results, and tie in with the local rules. If an individual exceeds an investigation level the employer must complete an investigation to determine the causes and ensure that adequate controls are in place to ensure that the individual's doses are ALARP.

Action levels linked to single monthly dosemeter readings are also given to help to highlight when there may be a potential problem, so that practices can be observed and optimisation strategies considered at an early stage. An action or investigation level of 2 mSv in a month based on the reading from a collar dosemeter has been proposed for interventional cardiologists (Durán *et al* 2013) and endorsed by the European Association of Percutaneous Cardiovascular Interventions and a number of other international societies, and this has been included here. Levels for other monitoring periods can be adjusted accordingly. These might be used in monitoring new staff or implementation of new techniques.

References

Behrens R and Dietze G 2011 Dose conversion coefficients for photon exposure of the human eye lens *Phys. Med. Biol.* **56** 415–37

Behrens R 2012 On the operational quantity Hp(3) for eye lens dosimetry *J. Radiol. Prot.* **32** 455–64

Bilski P *et al* 2011 The new EYE-D™ dosemeter for measurements of $H_p(3)$ for medical staff *Radiat. Meas.* **46** 1239–42

Cantone M C, Ginjaume M, Miljanic S, Martin C J, Akahane K, Mpete L, Michelin S C, Flannery C M, Dauer D T and Balter S 2017 Report of IRPA task group on the impact of the eye lens dose limits *J. Radiol. Prot.* **37** 527–50

Carinou E, Ferrari P, Bjelac O C, Gingaume M, Merce M S and O'Connor U 2015 Eye lens monitoring for interventional radiology personnel: dosemeters, calibration and practical aspects of $H_p(3)$ monitoring, A 2015 review *J. Radiol. Prot.* **35** R17–34

Clerinx P, Buls N, Bosmans H and de Mey J 2008 Double-dosimetry algorithm for workers in interventional radiology *Radiat. Prot. Dosimetry* **129** 321–7

Cousin A J, Lawdahl R B, Chakraborty D P and Koehler R E 1987 The case for radioprotective eyewear/facewear. Prqactical implications and suggestions *Invest. Radiol.* **22** 688–92

Domienik J, Bissinger A and Zmyslony M 2014 The impact of X-ray tube configuration on the eye lens and extremity doses received by cardiologists in electrophysiology room *J. Radiol. Prot.* **34** N73–9

Durán A, Hian S K, Miller D L, Le Heron J, Padovani R and Vano E 2013 Recommendations for occupational radiation protection in interventional cardiology *Catheter. Cardiovas. Interv.* **82** 29–42

Efstathopoulos E P, Pantos I, Andreou M, GKatis A, Carinou E, Koukorava C, Kelekis N L and Brounttzos E 2011 Occupational radiation doses to the extremities and the eys in interventional radiology and cardiology *Br. J. Radiol.* **84** 70–7

European Commission 2013 Council Directive 2013/59/Euratom of 5 December 2013 laying down basic safety standards for protection against the dangers arising from exposure to ionising radiation, and repealing Directives 89/618/Euratom, 90/641/Euratom, 96/29/Euratom, 97/43/Euratom and 2003/122/Euratom *Official J. Eur. Union* **57** L13 http://eur-lex.europa.eu/LexUriServ/LexUriServ.do?uri=OJ:L:2014:013:0001:0073:EN:PDF (assessed 26/4/16)

Farah J, Struelens L, Dabin J, Koukorava C, Donadille L, Jacob S, Schneizer M, Auvinen A, Vanahavere F and Clairand I 2013 A correlation study of eye lens dose and personal dose equivalent for interventional cardiologists *Radiat. Prot. Dosimetry* **157** 561–9

Ferrari P, Becker F, Carinou E, Chumak V, Farah J, Jovanovic Z, Krstic D, Morgun A, Principi S and Teles P 2016 Monte Carlo study of the scattered radiation field near the eyes of the operator in interventional procedures *J. Radiol. Prot.* **36** 902–21

Fetterly K A, Schueler B A, Grams M P and Sturchio G M 2017 Estimating head and neck tissue dose from X-ray scatter to physicians performing X-ray guided cardiovascular procedures: a phantom study *J. Radiol. Prot.* **37** 43–58

Franken Y 2002 Guidance on the use of protective lead aprons in medical radiology protection efficiency and correction factors for personal dosimetry *Proc. 6th European ALARA Network Workshop* (Madrid: Occupational Exposure Optimisation in the Medical Field and Radiopharmaceutical Industry), pp 135–9

Hafez M A, Smith R M, Matthews S J, Kalap G and Shermann K P 2005 Radiation exposure to the hands of orthopaedic surgeons: are we underestimating the risk? *Arch. Orthop. Trauma Surg.* **125** 330–5

Higgins A 2016 Measurement of occupational doses of ionising radiation to the lens of the eyes of interventional radiologists *J. Radiol. Prot.* **35** 74-92

Hu P, Kong Y, Chen B, liu Q, Zhuo W and Liu H 2017 Shielding effect of lead glasses on radiologists' eye lens exposure in interventional procedures *Radiat. Prot. Dosimetry* **174** 136–40

ICRP 2000 Pregnancy and medical radiation. ICRP Publication 84 *Ann. ICRP* **30** 33–4

ICRP 2007 The 2007 recommendations of the International Commission on Radiological Protection. ICRP Publication 103 *Ann. ICRP* **37**

ICRP 2013 Radiological protection in cardiology. ICRP Publication 120 *Ann. ICRP* **42**

ICRP 2018 Occupational radiological protection in interventional procedures. ICRP Publication 139 *Ann. ICRP* **47**

IRSN 2018 *Le dosime' tre cristallin DOSIRIS* (Croissy Sur Seine: Institut de Radioprotection, Surete´ Nucleaire) Available at: http://dosimetre.irsn.fr/fr-fr/Documents/Fiches%20produits/IRSN_Fiche_dosimetre_Cristallin.pdf (accessed 24/2/18)

ISO 2015 *Radiological Protection—Procedures for Monitoring the Dose to the Lens of the Eye, the Skin and the Extremities 15382:2015* (Geneva: ISO)

Järvinen H, Buls N, Clerinx P, Jansen J, Miljanić S, Nikodemová D, Ranogajec-Komor M and d'Errico F 2008a Overview of double dosimetry procedures for the determination of the effective dose to the interventional radiology staff *Radiat. Prot. Dosimetry* **129** 333–9

Järvinen H, Buls N, Clerinx P, Miljanić S, Nikodemová D, Ranogajec-Komor M, Streulens L and d'Errico F 2008b Comparison of double dosimetry algorithms for estimating the effective dose in occupational dosimetry of interventional radiology staff *Radiat. Prot. Dosimetry* **131** 80–6

Kicken P J H, Kemerink G J, Schultz F W, Zoetelief J, Broerse J J and van Engelshoven J M A 1999 Dosimetry of occupationally exposed persons in diagnostic and interventional arteriography. Part II: Assessment of effective dose *Radiat. Prot. Dosimetry* **82** 105–14

Kopeć R, Bubak A, Budzanowski M, Sas-Bieniarz A and Szumska A 2016 How do hospoital sterilisation procedures affect the response of personal extremity rings and of eye TL dosemeters? *Radiat. Prot. Dosimetry* **170** 302–6

Magee J S, Martin C J, Sandblom V, Carter M J, Almén A, Cederblad A, Jonasson P and Lundh C 2014 Derivation and appliaction of dose reduction factors for protective eyewear in interventional radiology and cardiology *J. Radiol. Prot.* **34** 811–23

Martin C J 2009 A review of radiology staff doses and dose monitoring requirements *Radiat. Prot. Dosimetry* **136** 140–57

Martin C J 2012 Personal dosimetry in the UK: is it time for a change? *J. Radiol. Prot.* **32** E3–6

Martin C J 2018 Protecting interventional radiology and cardiology staff: Are current designs of lead glasses and eye dosemeters fit for purpose? *J. Radiol. Prot.* **38** E22–5

Martin C J and Magee J S 2013 Assessment of eye and body dose for interventional radiologists, cardiologists, and other interventional staff *J. Radiol. Prot.* **33** 445–60

McCallum S, Watt M, Latif N, Redford C and Hampson L 2015 *Role out of Collar badges in NHS Grampian and NHS Forth Valley* https://www.ipem.ac.uk/Portals/0/Documents/Conferences/2015/315%20RPA%20Update/Steve%20McCallum.pdf (accessed 27/2/2018)

McVey S, Sandison A and Sutton D G 2013 An assessment of lead eyewear in interventional radiology *J. Radiol. Prot.* **33** 647–59

NCRP 2011 Radiation dose management for fluoroscopically guided interventional medical procedures *NCRP Report* **168**

PHE 2014 *Eye Dosemeters* (Oxford: Public Health England) https://www.phe-protectionservices.org.uk/cms/assets/gfx/content/resource_2974cs6788076842.pdf (accessed 25/2/2018)

Principi S, Ginjaume M, Duch A M, Sánchez R M, Fernández J M and Vano E 2015 Influence of dosemeter position for the assessment of eye lens dose during interventional cardiology *Radiat. Prot. Dosimetry* **164** 79–83

Siiskonen T, Tapiovaara M, Kosunen A, Lehtinen M and Vartiainen E 2007 Monte Carlo simulations of occupational radiation doses in interventional radiology *Br. J. Radiol.* **80** 460–8

Silva E H, Struelens L, Covens P, Ueno S, Ube M, Vanhavere F and Buls N 2017 Where is the best position to place a dosemeter in order to assess the eye lens dose when lead glasses are used? *Radiat. Meas.* **106** 257–61

Swiss Ordinance 2008 *Eidgenössisches Departement des Inneren und Eidgenössisches Departement für Umwelt, Verkehr, Energie und Kommunikation:* Verordhung über die Personendosimetrie (Dosimetrieverordnung) vom 07.10.1999 (Vers. 1 January 2008); SR814.501.43. Bern; 2008 (in German).

Vanhavere F *et al* 2012 ORAMED: Optimisation of Radiation Protection for Medical Staff *7th EURADOS Report* 2012-02 Available at: http://www.eurados.org./en/Documents_Publications (accessed May 2017)

Van Rooijen B D, de Hann M W, Das M, Arnoldussen C W K P, de Graaf R, van Zwam W H, Backes W H and Jeukens C R L P N 2014 Efficacy of radiation safety glasses in interventional radiology *Cardiovasc. Interv. Radiol.* **37** 1149–55 von Boetticher H, Lachmund J, Looe H K, Hoffmann W and Poppe B 2008 2007 Recommendations of the ICRP change basis for estimation of the effective dose: What is the impact on radiation dose assessment *Fortschr. Röntgenstr.* **180** 391–5

von Boetticher H, Lachmund J and Hoffmann W 2009 Cardiac catheterization: impact of face and neck shielding on new estimates of effective dose *Health Phys.* **97** 622–7

Von Boetticher H, Lachmund J and Hoffmann W 2010 An analytic approach to double dosimetry algorithms in occupational dosimetry using energy dependent organ dose conversion coefficients *Health Phys.* **99** 800–5

Chapter 5

Use of x-rays in diagnostic and interventional radiology

5.1 Introduction

A wide range of staff in healthcare are involved in some capacity in the use of x-rays. Apart from radiologists and radiographers, there are staff from a number of surgical specialties, anaesthetists, nurses, dentists and various allied health professionals, the majority of whom are unlikely to receive occupational exposures approaching or exceeding any relevant dose limits or constraints. However, hospital management, as advised by their radiation protection advisers, must ensure that risk assessments are carried out and decide upon who should be monitored and what dosemeters they should wear. Reviews of dose levels reported in the literature are provided in order to allow assessments of the likelihood of high exposures occurring in different applications to be assessed. The first part of this chapter deals with staff undertaking and assisting in diagnostic procedures in radiology departments, many of whom will be behind protective screens for most, if not all, of the time when exposures are being made. The second part deals with interventional radiology and cardiology, which involve operators and assistants being in close attendance to the patient during x-ray procedures, so they form some of the most highly exposed staff groups. The third part considers issues relating to groups using fluoroscopic equipment, particularly mobile C-arms, undertaking procedures that require images for guiding surgical intervention usually in theatre settings. These procedures are similar in nature to the interventional ones, requiring staff to be present in the controlled area when x-rays are being emitted, but they are generally less complex. However, because they are not necessarily carried out in specialised radiation facilities with the same range of protective devices, the doses received can be substantial.

doi:10.1088/2053-2563/aaddb6ch5

The recommendations given in the following subsections are for **non-classified** workers only. Recommendations given in section 4.6.2 should be followed for classified workers.

5.2 Diagnostic radiography and fluoroscopy in x-ray departments

Radiographers perform many different types of x-ray procedure. Their doses are likely to be below the detection limit of any dosemeter when they are located behind a protective screen, where their whole body and head will be protected. If this makes up the majority of their work then any dose received will be extremely small. However, many radiographers will sometimes assist in fluoroscopic or other examinations for which they will be present in the controlled area. It is at these times they are more likely to receive an exposure, but they will on these occasions be wearing a protective lead apron. In the past these staff have been monitored using a dosemeter often worn on their chest. When staff wore a lead apron this dosemeter would be under the apron. It was used to provide a record of their body dose, which was linked to effective dose in the dose record. However, the dose recorded will be close to the minimum level for the majority of the time, and is unlikely ever to reach levels of a few mSv. A result of the introduction of a 20 mSv dose limit for the lens of the eye with a 15 mSv classification level is that the doses to the eyes, which are not protected when the staff member is in a controlled area, are more likely to approach the limit for any staff working with x-rays. Therefore it becomes more important to record a measurement that can give an indication of the dose to the eyes than to the body.

> For staff involved in radiography (and not routinely wearing a dedicated eye dosemeter), a dosemeter worn at the collar outside any body protection can be used to record $H_p(3)$ (or $H_p(0.07)$) for assessment of exposure of the eyes. The $H_p(10)$ reading of the dosemeter may be used to give an indication of effective dose, and equation (4.2) be employed if staff wear a lead apron whenever they are working close to the x-ray source. However, if a lead apron is not always worn during periods of exposure, then an under-apron dosemeter may also be required to assess effective dose.

Since radiographers taking radiographs will in general be several metres from the x-ray sources and patients, the exposure of the eye is likely to be similar to that to the neck. A collar dosemeter will provide a measure of the radiation field to which the head and upper body of any staff member are exposed, and these levels can be used in risk assessments, based on which radiation protection measures can be planned. All radiographers and radiologists should be monitored initially and the majority will probably require regular monitoring. If they are likely to be present in any room where x-rays are being used, even if their dosemeters record minimal doses, it is recommended that they should continue to use a dosemeter, since higher levels of exposure may still occur in unusual incident situations. Additional dosemeters are unlikely to be required, but this can be assessed based on values of $H_p(10)$ recorded by the collar dosemeter using the approach set out in figure 4.2 and values in table 4.2. However, there are certain areas where the dose levels are minimal and the potential for exposure is so low that routine monitoring may not be appropriate and some discussion of these areas is included in the following subsections.

5.2.1 Mammography

Mammographic imaging involves the use of lower energy x-rays than in other x-ray modalities (25–35 kVp for conventional 2D imaging). The radiation levels are low because the thicknesses of tissue to be penetrated are small, and since the primary beam is collimated to the detector only scattered radiation is of concern. A suitable worst case level of scatter air kerma is 7.6 μGy per image at 1 m from the image receptor assessed for conventional mammography (Sutton *et al* 2012) and this value can be used to calculate scatter air kerma for all angles of scatter and any target/filter combination for the purpose of predicting scatter levels.

Mammographic imaging is typically carried out with a single operator positioned behind a protective lead screen. Provided a site employs good radiation safety management the likelihood of significant occupational exposure to staff operating mammography equipment is very low and staff doses are likely to be minimal even for the busiest departments. A review of personal dosimetry records from a busy dedicated breast screening centre at which all members of staff wore unprotected dosemeters for three monthly periods showed that of 133 dosemeters issued, results for 96 were below the minimum detectable dose (<0.01 mSv) and the maximum three monthly dose was 0.09 mSv (Jupp 2017).

One issue that should be taken into account is the small footprint of many mammography x-ray rooms, such that tertiary scatter levels from the walls, floor and ceiling may be greater. X-ray rooms on mobile vans may also have small footprints. In addition, the introduction of digital breast tomosynthesis has led to small increases in tube potentials (up to approximately 49 kVp) and patient doses, but this is unlikely to have a significant impact on protection considerations or the approach to staff dosimetry in mammography.

> For staff working in dedicated mammography centres (including those working in mobile vans) who are not exposed to other forms of ionising radiation, a trial period of wearing a collar dosemeter would provide information on eye and effective dose levels for full-time staff.

Results of these measurements will determine whether routine monitoring is necessary and inform risk assessments. If staff members are already issued with personal dosemeters because of work with other modalities, dosemeters should continue to be worn when working in mammography. For dedicated mammography staff who never wear lead aprons, the $H_p(10)$ reading will give the effective dose.

5.2.2 Mobile radiography

Radiographers are required frequently to perform x-rays on patients being treated in hospital wards using mobile equipment. They will normally wear lead aprons, but need to observe the patient during the exposure, so their eyes will be exposed to scattered radiation. The majority of ward radiography is for chest examinations for which the exposure levels are not high, so radiographers are unlikely to receive significant radiation doses.

> All staff taking mobile radiographs regularly are recommended to wear collar dosemeters to assess eye and effective dose. Corrections for PPE can be applied accordingly as described in chapter 4.

Dose levels to which staff may be exposed can be estimated from the kerma-area product (KAP) workload for use in risk assessments (see section 7.2.3), as well as determined from the collar dosemeters worn.

5.2.3 Dental radiography

Intra-oral dental radiography, which is performed in dental surgeries throughout the UK, is the x-ray examination carried out most frequently (Gulson *et al* 2007). For intra-oral exposures, the weighted average primary plus scatter dose level is approximately 0.5 μGy at 1 m from the patient (Worrall *et al* 2012, Sutton *et al* 2012). Dental professionals taking radiographs should be able to stand at over 2 m from the patient during an exposure, so the dose received should be only 0.1 μGy per radiograph. Therefore, a dentist taking 100 radiographs per week would be unlikely to exceed 0.5 mSv per year to either the body or eye. Panoramic radiology units involve exposure from different directions and the scatter level at 1 m is approximately 0.65 μGy per examination (Sutton *et al* 2012). Although, the operator can stand at some distance, he/she needs to observe the complete examination during which the tube rotates around the patient.

Existing national guidance (NRPB 2001) recommends personal monitoring should be considered for staff whose weekly workload exceeds 100 intra-oral or 50 panoramic images, or a pro rata combination of each type. Periodic monitoring at lower workloads can be useful to check that staff adhere to recommended protection practices and to reassure staff. The type of monitoring recommended is use of a dosemeter worn on the chest or at the collar. As staff do not normally wear lead aprons, the collar dosemeter measurements will provide direct measures of dose to the eye $H_p(3)$ and effective dose to the body $H_p(10)$.

> Initial monitoring of dental staff with a collar dosemeter is recommended to confirm that protection arrangements are adequate if the number of intra-oral radiographs taken per week is likely to exceed 100, the number of panoramic examinations exceeds 50 per week, or the combined workload exceeds the pro rata total. The collar dosemeter will provide dosimetry data that can be used in risk assessments, and based on the results and the workload it may be possible to discontinue routine monitoring in many clinics.

Doses from cone beam CT are of the order of 20–40 μGy at 1 m (Holroyd and Walker 2010). Dental staff using panoramic or cone beam equipment may need to wear a lead apron or use a screen to protect the body.

> Dental staff operating cone beam CT dental x-ray equipment are recommended to wear a collar dosemeter initially as a means to establish effective dose and eye dose levels to determine whether regular monitoring is required.

5.2.4 Dual energy x-ray absorptiometry (DEXA)

DEXA scanning is used for the assessment of bone densitometry and bone composition. The effective doses to the patient are generally less than 10 μSv. A risk assessment based on manufacturer scatter data (or measured dose rates) should be performed to confirm that dose levels in the facility are below the level at which dose monitoring is required (table 4.2).

> Dose monitoring should not normally be required for staff working exclusively with DEXA scanners.

5.3 Interventional radiology and cardiology

Interventional radiologists and cardiologists are some of the most highly exposed groups of healthcare staff. They perform complex procedures that involve manipulations under x-ray guidance, while viewing images on a monitor, and so need to be not only present in the x-ray room, but also close to the patient while x-rays are being emitted. Therefore there is the potential for them to receive high doses to their eyes, torso and extremities as they guide and manipulate needles and catheters while viewing the procedure using fluoroscopy (ICRP 2000, 2013).

The interventional operator stands at the side of the patient couch, more frequently on the right side of the patient, with the x-ray tube and gantry on their left. The position of the operator and their proximity to the x-ray tube have a strong influence on the amount of scatter to which they are exposed and therefore the doses recorded. Cardiologists performing percutaneous coronary interventions (PCIs) gain

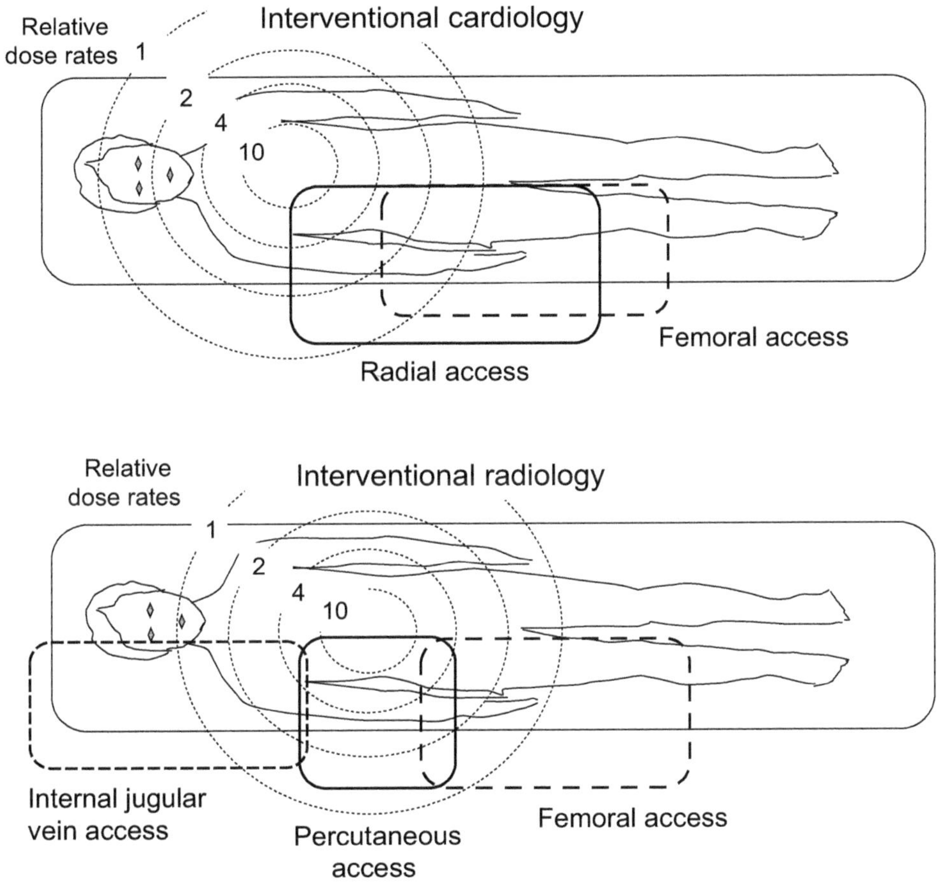

Figure 5.1. Positions where the hands of operators will be manipulating catheters during interventional cardiology using radial and femoral access routes, and interventional radiology procedures through internal jugular vein, percutaneous and femoral access. (Figure 16.3 in Martin and Sutton (2015), reproduced by permission of Oxford University Press https://global.oup.com/academic/.)

vascular access via either the radial artery in the hand or the femoral artery at the groin (figure 5.1). Procedures involving the radial access route require the operator to stand closer to the x-ray tube leading to higher doses, especially to the operator's hands and eyes, than those using femoral access. Radial access, however, often has advantages relating to treatment of the patient. Interventional radiologists frequently use the femoral artery and sometimes the internal jugular vein, but other procedures using percutaneous access require them to be closer to the x-ray tube where scatter levels are greater (figure 5.1). Percutaneous interventions may also be visualised through CT fluoroscopy which provides immediate feedback through sectional images. The operator is exposed to a considerable amount of scattered radiation while standing next to the patient. Operators performing any of these procedures need to take appropriate precautions, such as optimising the exposure conditions, utilising external protective devices, and wearing personal protective equipment such as lead aprons, thyroid shields, protective eyewear and sometimes protective gloves (Hiles *et al* 2016). The risk to the thyroid is high for females under 20 years, but falls to minimal levels by the age of 30 years and is negligible by the age of 40 years (ICRP 2007), and is lower for males than for females, so protection of the thyroid is particularly important for younger female staff present during interventional procedures. Dose monitoring is essential in order to ensure operators are using protective devices effectively and not approaching any dose limits (Martin 2016b).

5.3.1 Body dose monitoring

There have been a number of compilations of data from studies of doses to interventional clinicians (Kicken *et al* 1999, Kim *et al* 2008, Martin 2009, Miller *et al* 2010, Kuipers *et al* 2010, Padovani *et al* 2011, Kim *et al* 2012), and these show that annual effective doses are often within a range of 1–4 mSv, although the majority of recorded values are lower.

> Interventional clinicians are recommended to wear two dosemeters one preferably on the head adjacent to the eye (or at the collar above the lead apron) and one on the body underneath the apron, as this provides the best option for assessment of eye and effective dose (section 4.5).
>
> All non-classified staff working in an interventional suite are recommended to wear a dosemeter at the collar outside the lead apron. If the $H_p(10)$ reading exceeds 6 mSv ($E \approx 1$ mSv), then a dosemeter under the lead apron should be worn and the double dosimetry system adopted (figure 4.2 and table 4.2).

5.3.2 Eye dose monitoring

5.3.2.1 Interventional procedures

Interventional cardiologists and radiologists stand adjacent to the patient, with the x-ray tube and gantry to their side. The position of the monitor and the use of ceiling suspended screens and other protective devices will have a significant effect on the dose to the eye lens. There have been numerous studies of eye dose levels from interventional procedures, with typical average values of 20–100 μSv per procedure and 0.2–1.0 μSv per Gy cm^2 KAP (Lie *et al* 2008, Vanhavere *et al* 2012, Higgins 2016, Sánchez *et al* 2016). The dose to the eye for PCIs is typically in the region of 60 μSv per procedure. These levels can give monthly doses of 0.5–2 mSv to the eye, so there are risks that the eye dose limit can be exceeded if appropriate protection measures are not in place. However dose levels vary depending on the proximity of the operator to the irradiated area. In the European ORAMED study dose measurements were performed at multiple centres for three cardiac and five interventional radiology procedures (Vanhavere *et al* 2012). Operators received higher doses from embolization procedures (median 80 μSv per procedure, third quartile 180 μSv per procedure). Whereas during cerebral and carotid procedures using femoral access where the operator stands further from the region being irradiated the doses were approximately 40 μSv per procedure.

5.3.2.2 CT interventional techniques

Clinicians performing CT guided interventions can receive significantly greater doses to the eye lens. The use of several types of PPE is usually required and additional shielding adjacent to the scan plane may also be considered (Neeman *et al* 2006). Estimates of dose rates to the operator from CT fluoroscopy drop rapidly with distance from the scan plane. Measurements of scatter dose rates from phantoms have given values of 4 mGy s^{-1} in the scan plane, 9 μGy s^{-1} at the body trunk and 2 μGy s^{-1} to the eyes with the operator positioned 40 cm from the scan plane (Keat 2001, Nickoloff *et al* 2000). Doses of 20–40 μSv to the eye per minute of exposure have been reported during CT fluoroscopy with doses per procedure being similar to those for other interventional procedures (Saidatul *et al* 2010). However, dose levels are difficult to predict as they are dependent on the position and technique of the operator, and a wide range of dose levels per procedure recorded outside the lead apron have been reported ranging from 5–7 μSv (Teeuwisse *et al* 2001) up to 240 μSv (Buls *et al* 2003). Therefore particular care should be taken in monitoring doses to the eyes to ensure protection is optimised and the dose limit is adhered to.

5.3.2.3 Eye dose monitoring recommendations

The superficial dose $H_p(3)$ (or $H_p(0.07)$) to a dosemeter worn at the collar provides an indication of eye dose, rather than an accurate measurement.

> Interventional clinicians who record annual doses of 10 mSv on their collar dosemeter should wear an approved dedicated eye dosemeter to determine whether further control measures should be implemented and whether classification is necessary.

The protection provided by lead glasses depends on the angle at which scatter from the patient is incident on the head (McVey *et al* 2013, Van Rooijen *et al* 2014, Magee *et al* 2014). For the majority of the time that an interventional radiologist or cardiologist is carrying out a procedure, he/she will not be looking towards the patient when x-rays are being emitted. Therefore, the protection factor required must take account of x-ray beams incident from the side and below the level of the head (sections 4.3 and 4.6.4.2).

> The application of a protection factor of 0.5 to the eye dose is recommended when lead glasses of standard design are used. A lower factor between 0.3 and 0.4 based on measurements may be used, but the derivation must be documented. Details on the application of correction factors is given in section 4.6.4. Correction factors cannot be applied to the dose records by an ADS until the process has been approved by HSE.
>
> If any factor is applied the employer must be certain that the interventional clinician concerned always wears both the protective eyewear and the dosemeter, and can achieve this through implementation of a robust documented quality assurance programme, including regular checks to confirm compliance.

5.3.3 Hand monitoring

5.3.3.1 Interventional procedures

Interventional radiologists and cardiologists manipulate catheters aided by x-ray imaging, so their hands are close to the x-ray field. Hand doses per procedure in interventional radiology and cardiology vary from less than 10 μSv up to 2 mSv depending on the position of the operator and the complexity of the procedure (Felmlee *et al* 1991, Whitby and Martin 2005, Sauren *et al* 2011). In the multi-centre study by Vanhavere *et al* (2012) median doses for individual procedures of different types were in the range 20–80 μSv, and third quartile values between 60 and 280 μSv, although doses up to 7 mSv for an individual procedure were reported.

For the majority of interventional fluoroscopy procedures the hand is side-on to the x-ray beam and the outer aspect tends to receive a higher dose, so the best place to wear the dosemeter is often on the little finger of the hand closest to the x-ray tube (Whitby and Martin 2005, Vanhavere *et al* 2012). However, some thought should be given to the positions of the hands adopted by the individuals to be monitored and observation of techniques can provide valuable information on this.

Access to the arterial system can be gained via various routes (figure 5.1). For cardiology the femoral artery at the groin or the radial artery are used. The radial approach tends to give higher doses to the hands because of proximity to the area being imaged. The femoral artery is the more common access route for both interventional cardiology and radiology and makes up about 80% of procedures performed in interventional radiology. Illustrations of the hand positions for different access routes are shown in figure 5.2. It is usually the non-dominant (left) hand of interventional clinicians that is closer to the x-ray field, as this often holds the catheter in position, therefore the outer aspect of the left hand tends to receive the highest dose during femoral access (figures 5.2(a) and (b)) (Whitby and Martin 2005). The dominant (right) hand manipulates the inserted device using a combination of prodding and twisting actions. The pattern is similar for procedures using internal jugular vein access such as transjugular intrahepatic portosystemic shunt (TIPS) (figures 5.2(c) and (d)). For procedures using a percutaneous access route closer to the x-ray tube and image receptor (figures 5.2(e) and (f)), the movement of the hand is predominantly a prodding action, so that the tips of the index and middle fingers receive slightly higher doses than other parts of the hand.

> Wearing of a ring dosemeter is recommended if the finger dose is in the range 2–4 mSv per month. For the majority of interventional clinicians, wearing the dosemeter on the little finger of the hand nearest to the x-ray tube provides the best monitoring option. If the dose exceeds 4 mSv per month, the dose distribution across the hand should be determined and monitoring arrangements adjusted to take this into account.

If the workload of an individual is predominantly percutaneous procedures, monitoring of the fingertip might be considered. However, fingertip dosemeters

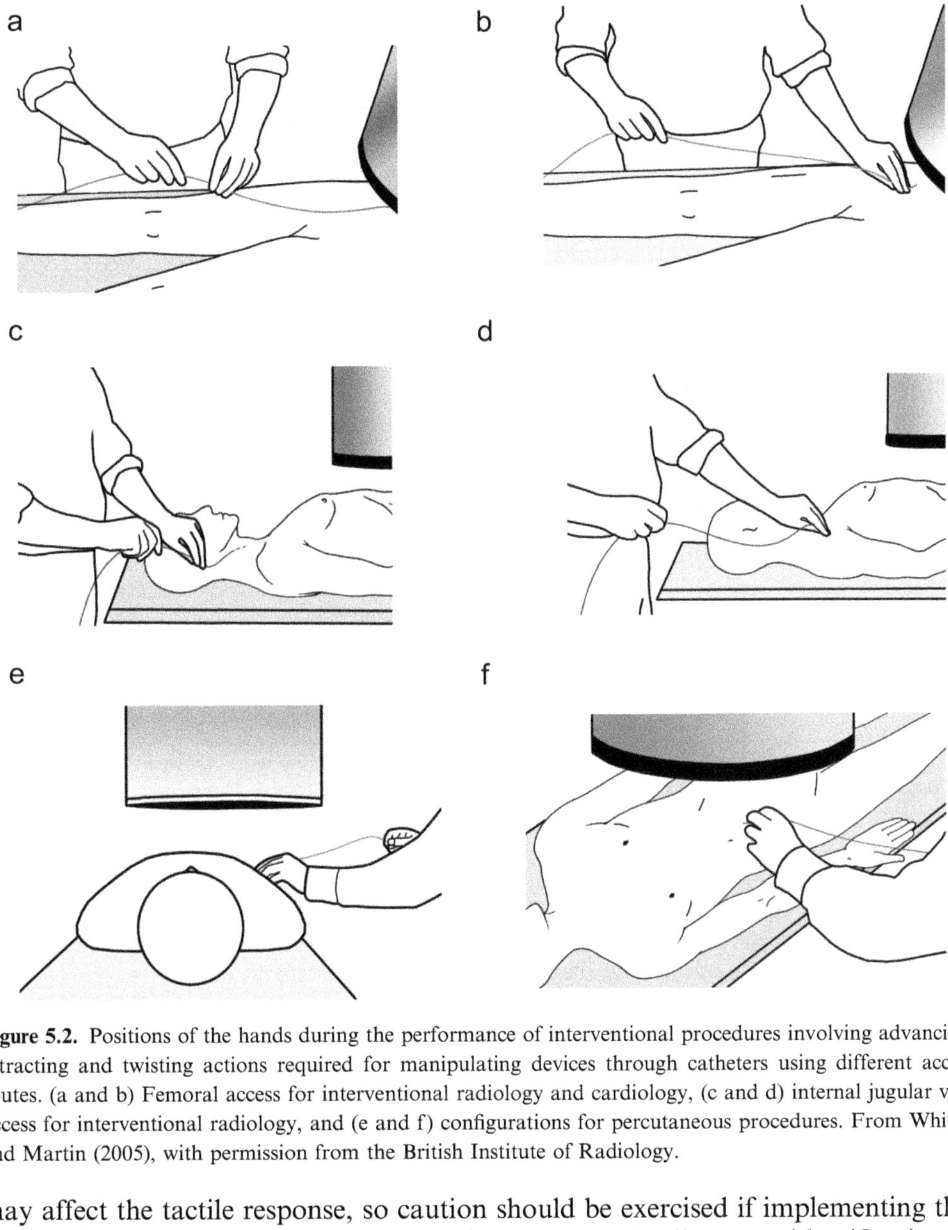

Figure 5.2. Positions of the hands during the performance of interventional procedures involving advancing, retracting and twisting actions required for manipulating devices through catheters using different access routes. (a and b) Femoral access for interventional radiology and cardiology, (c and d) internal jugular vein access for interventional radiology, and (e and f) configurations for percutaneous procedures. From Whitby and Martin (2005), with permission from the British Institute of Radiology.

may affect the tactile response, so caution should be exercised if implementing this approach. If both hands are monitored, then a system to allow easy identification of left and right dosemeters should be in place.

5.3.3.2 CT interventional techniques

The risk of the operator's hands or fingers entering the primary x-ray field is significant in CT fluoroscopy or 'step and shoot' interventional CT. Those performing CT interventional procedures may be required to stand close to the patient with their hands within a few centimetres of the primary beam, judging their position by means of the laser markers. A primary beam from a CT scanner can deliver doses of hundreds of μGy in a few seconds. Poor technique can result in operators' hands being placed in the direct beam, where the annual dose limit of 500 mSv could be reached in a few minutes (Buls *et al* 2003, Saidatul *et al* 2010). However, because the primary beams are relatively narrow (~5 mm wide), the chances of a significant dose received being recorded by a ring dosemeter worn at the base of the finger, or even a finger stall worn at the tip, are fairly low, and in most cases only scattered radiation would be measured (which would be significantly outweighed by primary beam exposure). Extremity monitoring in CT fluoroscopy is therefore of limited value and may provide false reassurance.

The CT images should alert an operator if part of their hand enters the primary beam. However, in one example of a CT fluoroscopy incident involving hand exposure, the collimation (irradiated beam width) was set to 14.4 mm, giving the radiologist three images of 4.8 mm. In this case the radiologist observed the image from the central beam during the procedure, and exposure of the hand was revealed only later when all the images were reviewed (Davis 2016). The capability to review images provides a facility for exercising control and facilitating investigations, so that better assessments of doses received can be made through measurements in the actual positions where exposures have occurred for appropriate lengths of time. Since dose levels in the CT beam can be measured using ionisation chambers or TLDs, dose data should also be included in risk assessments.

Personal monitoring of finger doses is generally not useful for CT fluoroscopy. A radiation risk assessment should be undertaken including measurements to determine likely dose levels if fingers were to enter the beam and training should be given to operators to ensure they are aware of risks and dose levels. Strict dose management arrangements involving review of images to ensure that the operator's hands have not entered the primary beam should be implemented. If dose monitoring is performed it could give false reassurance, so the limitations must be borne in mind.

5.3.4 Leg doses

During interventional radiology there is a need for the radiologist or cardiologist to stand close to the patient and x-ray tube in order to carry out complex procedures. For most interventional radiology procedures an undercouch C-arm configuration is used with the x-ray tube positioned below the couch. The primary beam incident on the patient is scattered in all directions. Upwards scatter is attenuated by passage through the patient whereas scatter downwards and to the side is predominantly unattenuated and therefore more intense (Artschan *et al* 2014). Scatter levels will be greatest when the x-ray beam is angled to irradiate the patient obliquely with the tube adjacent to the operator. Lead aprons are not designed to extend to the lower extremities, leaving these areas subject to potentially high levels of unattenuated scattered radiation from the patient. A lead/rubber curtain hung from the side of the patient couch is used to protect the lower extremities (Artschan *et al* 2014, Whitby and Martin 2003, Shortt *et al* 2007). These are often supplied as standard and are integral to the x-ray unit, but the lead thickness and hang length of the curtain vary between manufacturers. They should always be considered for interventional suites. Where no additional protective devices are used, doses to the legs and feet can be higher than those received by the hands (Whitby and Martin 2003, Vanhavere *et al* 2012). The multi-centre study by Vanhavere *et al* (2012) reported median doses to the leg of 20–60 μSv per procedure and third quartile values between 30 and 200 μSv.

A curtain with a lead equivalence of 0.25 mm Pb or greater is considered adequate to reduce doses to safe levels, for those regions located behind the shield and monitoring of these regions should not usually be required. However, consideration must be given to assess whether more distal regions of the lower extremities are fully shielded by the lead curtain (Hiles *et al* 2016). Depending on the hang length of the curtain and height of the patient couch, there may be a substantial gap between the bottom of the curtain and the floor. This may be a greater problem for taller interventional clinicians who need to operate with a higher patient couch. Therefore the feet and toes (which can protrude underneath the lead curtain), may not be fully protected and can be subject to high doses (Artschan *et al* 2014). In addition, for certain procedures such as TIPS, the radiologist is likely to stand at the head of the patient couch (figure 5.1) (Whitby and Martin 2003), and on some systems it may not be possible to use the lead curtain in this position. Where protective devices are not used, the dose to the leg can be as high as a few mSv per procedure (Whitby and Martin 2003).

> Where no lead curtain is used on an interventional couch, or the lead protection does not extend low enough to shield distal regions, a risk assessment should be carried out to determine skin doses to the lower extremities, and appropriate control measures put in place to ensure doses are ALARP. If residual doses to the shins and feet are likely to be of the order of 10 mGy per month or greater, dosemeters should be attached to the shoe or lower leg in order to assess levels and further protection measures implemented.

Practices should be observed to decide whether dose levels are likely to be an issue. The skin dose to unprotected regions of the lower extremities can be estimated using the KAP to skin dose conversion factor of 1 mGy/100 Gy cm^2 (Whitby and Martin 2003). Simple protective measures, such as taking a short step back while x-rays are being emitted, may be sufficient to reduce doses and it is unlikely that monitoring will need to be continued for an extended period.

Radiographers and ancillary staff such as nurses generally stand further from the patient and x-ray tube and therefore doses to the lower extremities are usually substantially lower than those of the radiologists or cardiologists. However, ancillary staff often stand on the opposite side of the patient where there is no protective curtain (Whitby and Martin 2003, Shortt *et al* 2007, Efstathopoulos *et al* 2011).

Doses to the legs of attendant staff who stand close to the x-ray source, but not behind a protective curtain can be significant and it is recommended that these are evaluated through risk assessment and dose monitoring if appropriate.

5.4 Mobile C-arm

Over the past few decades the introduction of mobile C-arm systems has led to an increased use of fluoroscopic x-ray imaging in a variety of departments within the hospital environment. Mobile C-arms are now commonplace in orthopaedic theatres, in gastroenterology often within dedicated endoscopy suites, in cardiology for pacemaker insertion, and used by vascular surgeons. Mobile C-arms may also be utilised in urology, gynaecology, pain clinics and by other specialties.

The radiation risk associated with fluoroscopic x-ray imaging is well known and personal protective equipment (PPE) in the form of a lead apron should be available to staff in all departments (Hiles *et al* 2016). However, additional PPE such as lead gloves and protective eyewear may not be used as frequently by staff working with mobile C-arm units when compared to their use in dedicated interventional suites or catherisation laboratories. Other protective devices such as overhead protective screens and lead curtains are also less likely to be available to staff.

The main source of occupational exposure associated with the use of mobile C-arm systems is from scattered radiation arising from the patient and operating table. However, there is also the potential for the hands to become exposed to direct radiation if they enter the primary beam. Although avoiding this is a well-established safety principal, a number of studies report higher extremity doses likely to be due to hands inadvertently straying into the beam (Kesavachandran *et al* 2012, Hafez *et al* 2005, Jones and Stoddart (1998), Müller *et al* 1998).

Because of the risk of exposure to the lens of the eye, the ICRP recommends that 'methods which provide reliable estimates of eye dose under practical situations should be established' where fluoroscopically guided procedures are performed outside the imaging department (ICRP 2012). Dosimetry data reported in this section show wide variations, presumably resulting from differences in technique, experience and observance of radiation protection principles, so dose monitoring is particularly important.

5.4.1 Orthopaedic medicine

Exposure of the orthopaedic surgeon depends on the type of procedure. Particular attention is given to intramedullary nailing and closed reduction procedures (Singer 2005, Levin *et al* 1987, Müller *et al* 1996, 1998, Mahaisavariya *et al* 2005), with several authors (Sanders *et al* 1993, Kesavachandran *et al* 2012, Mahajan *et al* 2015) reporting higher doses when compared to other open or percutaneous procedures. Surgery involving the femur (Kesavachandran *et al* 2012) and other larger body parts such as the spine (Rampersaud *et al* 2000) are also associated with higher exposures.

Several members of the surgical team are in the operating room for most orthopaedic procedures. This may include radiographers, nurses and anaesthetists, as well as any supporting surgeons assisting the primary orthopaedic surgeon. The highest doses are received by the members of staff closest to the tube and patient (Tasbas *et al* 2003, Palácio *et al* 2013), such as the primary and assisting surgeons, while exposures to the scrub nurse and anaesthetist positioned at greater distances are likely to be significantly lower (Mehlmann and DiPasquale 1997).

The experience of the orthopaedic surgeon also affects the level of occupational exposure, with less experienced orthopaedic registrars (Hafez *et al* 2005) and trainees (Botchu and Ravikumar 2008) receiving significantly higher exposures than experienced surgeons (Blattert *et al* 2004). Hafez *et al* (2005) also found a significant difference between the doses measured at the fingertip compared to those measured at the base of the index finger for the dosemeters worn by the trainee. The same was not seen for the consultant, which may suggest that the fingertips of less experienced surgeons are more likely to enter the primary beam.

In orthopaedic theatres the use of mini C-arms has become more widespread. These systems are designed to image smaller body parts such as hands, wrists and ankles, and therefore the amount of scattering material in the beam is low. This combined with lower energy exposure factors results in minimal scattered radiation (Giordano *et al* 2007, Singer 2005, Badman *et al* 2005, Giordano *et al* 2009, Mesbahi and Rouhani 2008). However, these systems may be used in an overcouch configuration in which doses can be higher, especially if the fingers enter the primary beam. Several studies have concluded that dose rates outside of the primary beam are low, and occupational exposures are significantly reduced when compared to conventional C-arm systems (Athwal *et al* 2005, Sung *et al* 2016, Dawe *et al* 2009, Giordano *et al* 2009). However, work carried out using mini C-arm systems is often intricate and there is a requirement for the surgeon to be positioned in close proximity to the patient and C-arm. Therefore, whilst the scattered radiation is small, occupational exposure should not be considered as zero (Giordano *et al* 2009, Sung *et al* 2016), particularly when the overcouch configuration is used and workloads are high (Domienik *et al* 2014).

5.4.1.1 Extremity exposure

Extremity doses reported in the literature vary considerably. A systematic review (Kesavachandran *et al* 2012) of 15 studies of radiation exposure to the hands/fingers of orthopaedic staff reported mean annual doses among orthopaedic surgeons and technical assistants ranging from 0.03 to 1.3 mSv, but more significant doses are seen in other studies. Goldstone *et al* (1993) investigated hand exposure during a mixture of internal and external fixation procedures for nine orthopaedic surgeons. The authors found that the exposure varied substantially between surgeons with monthly hand exposures ranging from 0.05–2.3 mSv. If the highest exposure from this study is extrapolated, the annual exposure becomes 28 mSv. Similar levels of exposure were found by Noordeen *et al* (1993) with extrapolated annual doses of 47.4 mSv for trauma surgeons. The highest extremity doses noted in the literature are observed for intramedullary nailing procedures, with estimated annual exposures reaching 72 mSv and 300 mSv (Tsalafoutas *et al* 2008, Müller *et al* 1998), although the later study concluded that the majority of radiation exposure was from hands entering the primary beam.

The majority of procedures will be carried out with the x-ray tube below the patient, so without additional shielding the legs may be at risk. However, the hands of the surgeons are likely to be closer to the x-ray beam, and so receive higher doses. In one of the few studies in which this was investigated, Tsalafoutas *et al* (2008) found the dose to the legs to be approximately half of that of the dose to the hands.

5.4.1.2 Eye lens exposure

Studies of eye doses have in the past been limited, but there are some studies which report eye exposures that are significant when compared with the new eye dose limit (Kesavachandran *et al* 2012). Earlier studies found extrapolated annual doses between 2 mSv and 12 mSv (Lo *et al* 1996, Harstall *et al* 2005) which are comparable to more recent studies with doses ranging from 4 mSv to 12 mSv (based on a 4 min per procedure, 250 procedures per year model) (Hoffler and Ilyas 2015, Romanova *et al* 2015, Tsalafoutas *et al* 2008). The need for careful monitoring of eye dose in orthopaedic medicine and the potential for staff to be designated as classified workers is clear.

5.4.2 Endoscopic medicine

Several studies have been undertaken investigating occupational exposure from endoscopic procedures and particular attention is given to Endoscopic Retrograde Cholangio-Pancreatography (ERCP) procedures (Naidu *et al* 2005, Buls *et al* 2002, Tsapaki *et al* 2011, Cohen *et al* 1997, O'Connor *et al* 2013, Zagorska *et al* 2015). Higher doses have also been associated with Percutaneous Transhepatic Cholangiography (PTC) procedures (Olgar *et al* 2009).

Occupational exposure depends on the type of fluoroscopic equipment being used, with higher doses to the hands and eyes being reported when an overcouch technique is employed (Dumonceau *et al* 2012, O'Connor *et al* 2013). Johlin *et al* (2002) also found significantly higher doses when a mobile C-arm was used compared to a fixed C-arm. Staff carrying out ERCP procedures in dedicated endoscopy suits tend to have more limited knowledge of radiation protection than their colleagues in the main radiology department (Buls *et al* 2002), and Vanhavere *et al* (2012) reported that in a multi-centre study 46% of those carrying out ERCPs did not use any room protective equipment. In addition, as for orthopaedic work, the availability of personal protective devices may be more limited when compared to interventional and cardiology suites.

5.4.2.1 Extremity exposure

Domienik *et al* (2011) included ERCP examinations in their study investigating extremity and eye lens doses in interventional radiology. Whilst this data was not specifically for mobile systems, the study demonstrated that extremity doses associated with ERCP procedures are lower than with other interventional procedures. Doses measured at the fingers, wrists and legs were found to be in the region of 20 μSv per procedure. This level of exposure is similar to that measured by Naidu *et al* (2005) who reported extrapolated annual extremity doses of 7.9 mGy. However, other authors have reported significantly higher extremity doses up to 0.4 mGy per ERCP procedure for an overcouch system (Buls *et al* 2002), and 0.8 mGy and 1 mGy per procedure for ERCP and PTC procedures, respectively, for an undercouch system (Olgar *et al* 2009).

5.4.2.2 Eye lens exposure

Eye lens doses reported in the literature appear more consistent than extremity doses. A detailed investigation by O'Connor *et al* (2013) reported annual doses to the left eye of the gastroenterologist of 1.3 mSv and 11.7 mSv for a fixed undercouch and a mobile overcouch system, respectively. Eye doses were significantly higher for the overcouch system despite the patient KAP being lower. The multi-centre study by Vanhavere *et al* (2012), which was not using mobile systems, reported a median dose of 18 μSv per procedure and a third quartile of 40 μSv, but the range was wide, with the maximum value being 4 mSv per procedure. Other studies report slightly higher levels of eye exposure from undercouch systems ranging from 94 μSv to 130 μSv per procedure for ERCP (Germanuad *et al* 1993, Olgar *et al* 2009) and 306 μSv per procedure for PTC (Olgar *et al* 2009). Buls *et al* (2002) report a median eye dose of 340 μSv per ERCP procedure using an overcouch tube. Vanhavere *et al* (2012) reported a median eye dose per unit KAP of 2 μSv $(\text{Gy cm}^2)^{-1}$ for ERCPs, although again measurements were not made on mobile systems.

A more recent study carried out by Zagorska *et al* (2015) assessed eye doses from ERCP to other members of staff using a mobile C-arm with the tube positioned underneath the patient. Doses were measured for the gastroenterologists, assisting staff, nurse and anaesthetist and annual extrapolated doses were estimated to be 19.4, 5.3, 8.1 and 28.3 mSv, respectively. The study showed the expected relationship between distance and dose, with the highest dose to the anaesthetist who stood closest to the x-ray beam. It demonstrates the potential for doses to approach and exceed the revised dose limit where protection has not been optimised.

5.4.3 Urology

Reported doses to the eye from percutaneous nephrolithotomy procedures show substantial variation with mean values of 34 μSv (Safak *et al* 2009), 180 μSv (Ritter *et al* 2012) and 300 μSv (Vano *et al* 2016) per procedure. The latter study reported median and third quartile doses per procedure of 243 μSv and 352 μSv respectively for the eye dose, and 394 μSv and 848 μSv for the collar dose, with maximum values over 1 mSv. If staff receive doses to the eye of this magnitude, they could readily exceed the 20 mSv annual dose limit, so monitoring is essential. Eye doses from other types of urology procedure range from 40 to 200 μSv per procedure (Ritter *et al* 2012, Taylor *et al* 2013).

Finger doses from urology are generally in the range 30–200 μSv per procedure, although Ritter *et al* (2012) report mean finger doses of 4.36 mSv for percutaneous nephrolithotomy procedures.

5.4.4 Pacemaker insertion

Permanent pacemaker implantation for bradycardia is carried out in large numbers of patients. An estimated 31 800 pacemaker implantations were performed in the UK in 2007 (Faulkner and Werduch 2008). Eye doses from pacemaker insertions and implantation of cardiac defibrillators are similar to those from coronary angiography. Vanhavere *et al* (2012) in a multi-centre study reported median and third quartile eye doses of 29 μSv and 62 μSv per procedure, with a maximum value of 1 mSv. The eye doses from these procedures are substantial because they are performed on mobile C-arm units which have less provision of specialised shielding devices, rather than being due to high patient doses. The mean eye dose per KAP was 5.5 μSv $(\text{Gy cm}^2)^{-1}$, which is approximately six times that found in interventional cardiology. Vanhavere *et al* also studied finger doses and reported median and third quartile values of 167 μSv and 405 μSv respectively per procedure, with a maximum value of 6.5 mSv.

5.4.5 Monitoring recommendations for orthopaedic, endoscopic, urology and other users of mobile C-arm systems

The studies reported in the literature show that there is the potential for high doses to the hands and eye lens of staff carrying out orthopaedic, endoscopic and urology procedures and to other users of mobile C-arm systems. High doses are reported using both the standard and inverted (x-ray tube above patient) C-arm orientation and exposure is shown to depend on the workload, type of procedures being carried out, and experience of the clinician. Although in some studies the maximum doses to the hands are higher, eye doses have a greater significance and should be considered the limiting factor for occupational exposure.

Mobile C-arms are utilised in other departments such as obstetrics/gynaecology and pain clinics, although data on occupational exposure is more limited. Given the recent reduction in the eye lens dose limit (IRR17) it is recommended that a collar dosemeter is used in the first instance to estimate occupational exposure of the operator and staff positioned close to the patient for all these groups. Initial monitoring is recommended for the operating clinician carrying out the procedure in all cases, but where workloads are low, an estimate of occupational exposure for staff assisting in the procedure can be performed using the patient KAP or by positioning a dosemeter on the C-arm itself, as described in section 7.2.

A whole body dosemeter worn underneath the protective apron is of little benefit and will not show any significant levels of exposure.

> All staff using mobile C-arm units are recommended to wear a collar dosemeter. Results from collar dosemeters should be used to estimate approximate eye doses and effective doses as described in chapter 4, and to determine whether additional monitoring is required from levels set out in figure 4.2 and table 4.2.
>
> Extremity monitoring may need to be considered if the operator's hands are close to the primary beam and high doses are recorded on collar dosemeters, particularly if mobile systems are used in the inverted orientation.
>
> Consideration should also be given to which members of the team require monitoring based on their proximity to the patient. The highest doses may not necessarily be to the clinician operating the equipment, as one might expect intuitively without reviewing the procedure (see section 5.4.2.2).

5.5 Training in dose optimisation

5.5.1 Education and training

Ensuring that interventional clinicians and in particular staff using mobile C-arm equipment wear their dosemeters can be challenging. Since the lives of their patients often depend on successful performance of the procedures, the wearing of dosemeters can be seen as a distracting and unnecessary imposition, as the risk of radiation effects to themselves is orders of magnitude less than the immediate risk to the patient from the procedure. Clinicians require relevant education and training about the potential effects of exposure, particularly on the eye to encourage compliance in wearing dosemeters and using protective devices (ICRP 2009). Employers will need to have systems in place to ensure compliance, including personal monitoring policies, efficient systems for the distribution, wearing and collection of dosemeters, clear instructions in local rules and periodic audits of dosemeter use. Disciplinary processes need to be considered to deal with issues of non-compliance.

5.5.2 Use of EPDs in optimisation

EPDs that provide immediate feedback to staff about cumulative doses and dose rates are useful in improving techniques and reducing doses to staff in fluoroscopically guided procedures (section 2.3). Systems that provide real-time visualisation of occupational radiation dose rate, intended for monitoring the personal exposure of medical staff working with fluoroscopically guided procedures, are now available commercially. The DoseAware EPD system, (Philips Healthcare, Best, The Netherlands) consists of individual dosemeters that record the radiation dose rate at 1 s intervals (Vañó *et al* 2011). The dosemeters are calibrated to measure $H_p(10)$ and are normally worn at chest height outside a lead apron. The radiation dose rate for several staff can be displayed on a monitor as a colour-coded bar chart on a logarithmic scale. The bars are green if the radiation dose rate is below 0.2 mSv h^{-1}, yellow if between 0.2 and 2 mSv h^{-1} and red if above 2 mSv h^{-1}. The radiation dose data are also stored to enable retrospective analysis.

Instant feedback and visualisation of radiation dose rates in this manner can raise awareness of the presence of ionising radiation, reminding staff to take dose-reducing actions. This can encourage individual responsibility for the use of radiation protection tools. For example it is instantly apparent from the rise in dose rate when a ceiling suspended screen or other protective device between the patient and the operator is moved. So the operator is alerted and can take immediate steps to reduce the dose rate to which he/she is exposed. Reductions in personal doses and a positive impact on optimisation of protection have been reported following the installation of systems of this type (Sanchez *et al* 2010, Chiriotti *et al* 2011, Clairand *et al* 2011, Sandblom *et al* 2013). Staff have a greater willingness to comply with use of protective devices when they are aware of their impact, so if staff recognise actions that lead to a reduction in dose they are more likely to cooperate in adoption of radiation protection measures. Purchase of such systems for interventional suites is encouraged.

Electronic dosemeters of this type could also be attached to the C-arm gantry to give an overall indication of when dose rates are high (Vañó *et al* 2013, 2016). This could be useful for other professionals such as anaesthetists, physiological measurement technicians and nurses who may not wear individual electronic monitors. Correlation of dose rates recorded by dosemeters worn by individual staff with those from a dosemeter attached to the C-arm gantry, which relate to the radiation level from the patient exposure, may be useful in determining whether effective use is being made of protective devices such as ceiling suspended screens. When there are concerns about dose levels being received by certain staff members, then it can be instructive to assign someone to observe procedures and record actions for a session while EPDs are in use. This is a role that medical physicists can usefully undertake as it allows them to observe practices at first hand. It is useful for trainee physicists, as they are able to carry out the analytical tasks at the same time as gaining an understanding of the clinical processes involved.

References

Artschan R E R, Brettle D S, Chase K, Fender A, Howells P G and Buchan S 2014 An investigation of the radiation doses to the lower legs and feet of staff undertaking interventional procedures *Br. J. Radiol.* **87** 20130746

Athwal G S, Bueno R A Jr and Wolfe S W 2005 Radiation exposure in hand surgery: mini versus standard C-arm *J. Hand Surg.* **30** 1310–6

Badman B L, Rill L, Butkovich B, Arreola M and Vander Griend R A 2005 Radiation exposure with use of the mini-C-arm for routine orthopaedic imaging procedures *J. Bone Joint Surg.* **87** 13–7

Blattert T R, Fill U A, Kunz E, Panzer W, Weckbach A and Regulla D F 2004 Skill dependence of radiation exposure for the orthopaedic surgeon during interlocking nailing of long-bone shaft fractures: a clinical study *Arch. Orthop. Trauma Surg.* **124** 659–64

Botchu R and Ravikumar K 2008 Radiation exposure from fluoroscopy during fixation of hip fracture and fracture of ankle: Effect of surgical experience *Indian J. Orthop.* **42** 471–3

Buls N, Pages J, Mana F and Osteaux M 2002 Patient and staff exposure during endoscopic retrograde cholangiopancreatography *Br. J. Radiol.* **75** 435–43

Buls N, Pagés J, de Mey J and Osteaux M 2003 Evaluation of patient and staff doses during various CT fluoroscopy guided interventions *Health Phys.* **85** 165–73

Chiriotti S, Ginjaume M, Vañó E, Sanchez R, Fernandez J M, Duch M A and Sempau J 2011 Performance of several active personal dosemeters in interventional radiology and cardiology *Radiat. Meas.* **46** 1266–70

Clairand I *et al* 2011 Use of active personal dosemeters in interventional radiology and cardiology: tests in laboratory conditions and recommendations—ORAMED project *Radiat. Meas.* **46** 1252–7

Cohen R V, Aldred M A, Paes W S, Fausto A M, Nucci J R, Yoshimura E M, Okuno E, Garcia M E, Maruta L M and Tolosa E M 1997 How safe is ERCP to the endoscopist? *Surg. Endosc.* **11** 615–7

Davis A 2016 private communication

Dawe E J, Fawzy E, Kaczynski J, Hassman P and Palmer S H 2009 A comparative study of radiation dose and screening time between mini C-arm and standard fluoroscopy in elective foot and ankle surgery *Foot Ankle Surg.* **17** 33–6

Domienik J *et al* 2011 Extremity and eye lens doses in interventional radiology and cardiology procedures: first results of the ORAMED project *Radiat. Prot. Dosimetry* **144** 442–7

Domienik J, Bissinger A and Zmyslony M 2014 The impact of X-ray tube configuration on the eye lens and extremity doses received by cardiologists in electrophysiology room *J. Radiol. Prot.* **34** N73–9

Dumonceau J M *et al* 2012 Radiation protection in digestive endoscopy: European Society of Digestive Endoscopy (ESGE) Guideline *Endoscopy* **44** 408–24

Efstathopoulos E P, Pantos I, Andreou M, GKatis A, Carinou E, Koukorava C, Kelekis N L and Brounttzos E 2011 Occupational radiation doses to the extremities and the eys in interventional radiology and cardiology *Br. J. Radiol.* **84** 70–7

Faulkner K and Werduch A 2008 Analysis of the frequency of interventional cardiology in various European countries *Radiat. Prot. Dosimetry* **129** 74–6

Felmlee J P, McGough P F, Morin R L and Classic K L 1991 Hand dose measurements in interventional radiology *Health Phys.* **60** 265–7

Germanaud J, Legoux J L, Sabattier R, Causse X and Trinh D H 1993 Radiation protection of operators during endoscopic retrograde cholangiopancreatography *Gastroenterol. Clin. Biol.* **1** 259–63

Giordano B D, Baumhauer J F, Morgan T L and Rechtine G R 2009 Patient and surgeon radiation exposure: comparison of standard and mini-C-arm fluoroscopy *J. Bone Joint Surg.* **91** 297-304

Giordano B D, Ryder S, Baumhauer J F and DiGiovanni B F 2007 Exposure to direct and scatter radiation with use of mini-C-arm fluoroscopy *J. Bone Joint Surg.* **89** 948–52

Goldstone K E, Wright I H and Cohen B 1993 Radiation exposure to the hands of orthopaedic surgeons during procedures under fluoroscopic X-ray control *Br. J. Radiol.* **66** 899–901

Gulson A D, Knapp T A and Ramsden P G 2007 *Doses to Patient Arising from Dental X-ray Examinations in the UK 2002–2004. A Review of X-ray Protection Service Data. HPARPD-022* (Chilton: HPA))

Hafez M A, Smith R M, Matthews S J, Kalap G and Shermann K P 2005 Radiation exposure to the hands of orthopaedic surgeons: are we underestimating the risk? *Arch. Orthop. Trauma Surg.* **125** 330–5

Harstall R, Heini P F, Mini R L and Orler R 2005 Radiation exposure to the surgeon during fluoroscopically assisted percutaneous vertebroplasty: a prospective study *Spine* **30** 1893–8

Higgins A 2016 Measurement of occupational doses of ionising radiation to the lens of the eyes of interventional radiologists *J. Radiol. Prot.* **35** 74–92

Hiles P A, Hughes H, Arthur D and Martin C J 2016 *Guidance on Personal Protective Equipment for Diagnostic X-ray Applications* (London: British Institute of Radiology)

Hoffler C E and Ilyas A M 2015 Fluoroscopic radiation exposure. Are we protecting ourselves adequately? *J. Bone Joint Surg. Am.* **97** 721–5

Holroyd J R and Walker A 2010 *Recommendations on the Design of X-ray Facilities and the Quality Assurance of Dental Cone Beam CT Systems. HPA-RPD-065* (Chilton: HPA)

ICRP 2000 Avoidance of radiation injuries from medical interventional procedures. ICRP Publication 85 *Ann. ICRP* **30**

ICRP 2007 The 2007 recommendations of the International Commission on Radiological Protection. ICRP Publication 103 *Ann. ICRP* **37**

ICRP 2009 Education and training in radiological protection for diagnostic and interventional procedures. ICRP Publication 113 *Ann. ICRP* **39**

ICRP 2012 Radiological protection in fluoroscopically guided procedures performed outside the imaging department. ICRP Publication 117 *Ann. ICRP* **40**

ICRP 2013 Radiological protection in cardiology. ICRP Publication 120 *Ann. ICRP* **42**

Johlin F C, Pelsang R E and Greenleaf M 2002 Phantom study to determine radiation exposure to medical personnel involved in ERCP fluoroscopy and its reduction through equipment and behavior modifications *Am. J. Gastroenterol.* **97** 893–7

Jones D J and Stoddart J 1998 Radiation use in the orthopaedic theatre: a prospective audit *Aust. N. Z. J. Surg.* **68** 782–4

Jupp T 2017 personal communication

Keat N 2001 Real-time CT and CT fluoroscopy *Br. J. Radiol.* **74** 1088–90

Kesavachandran C N, Haamann F and Nienhaus A 2012 Radiation exposure of eyes, thyroid gland and hands in orthopaedic staff: a systematic review *Eur. J. Med. Res.* **17** 28

Kicken P J H, Kemerink G J and van Engelshoven J M A 1999 Dosimetry of occupationally exposed persons in diagnostic and interventional arteriography. Part I: Assessment of entrance doses *Radiat. Prot. Dosimetry* **82** 93-103

Kim K P, Miller D L, Balter S, Kleinerman R A, Linet M S, Kwon D and Simon S L 2008 Occupational radiation doses to operators performing cardiac catheterization procedures *Health Phys.* **94** 211–27

Kim K P, Miller D L, Berrington de Gonzalez A, Balter S, Kleinerman R A, Ostroumova E, Simon S L and Linet M S 2012 Occupational radiation doses to operators performing fluoroscopically-guided procedures *Health Phys.* **103** 80–99

Kuipers G, Velders X L and Piek J J 2010 Exposure of cardiologists from interventional procedures *Radiat. Prot. Dosimetry* **140** 259–65

Levin P E, Schoen R W Jr. and Browner B D 1987 Radiation exposure to the surgeon during closed interlocking intramedullary nailing *J. Bone Joint Surg. Am.* **69** 761–6

Lie Ø Ø, Paulsen G U and Wøhni T 2008 Assessment of effective dose and dose to the lens of the eye for the interventional cardiologist *Radiat. Prot. Dosimetry* **132** 313–8

Lo N N, Goh P S and Khong K S 1996 Radiation dosage from use of the image intensifier in orthopaedic surgery *Singapore Med. J.* **37** 69–71

Magee J S, Martin C J, Sandblom V, Carter M J, Almén A, Cederblad A, Jonasson P and Lundh C 2014 Derivation and appliaction of dose reduction factors for protective eyewear in interventional radiology and cardiology *J. Radiol. Prot.* **34** 811–23

Mahaisavariya B, Songchareon P and Riansuwan K 2005 Radiation scattering to the primary surgeon during closed locked femoral nailing *J. Med. Assoc. Thai.* **88** 252–5

Mahajan A, Samuel S, Saran A K, Mahajan M K and Mam M K 2015 Occupational radiation exposure from C-arm fluoroscopy during common orthopaedic surgical procedures and its prevention *J. Clin. Diagn. Res.* **9** RC01–4

Martin C J 2009 A review of radiology staff doses and dose monitoring requirement *Radiat. Prot. Dosimetry* **136** 140–57

Martin C J 2016 Eye lens dosimetry for interventional radiologists and cardiologists: Practical approaches to protection and dose monitoring *Radiat. Prot. Dosimetry* **169** 286–91

Martin C J and Sutton D G 2015 Interventional radiology and cardiology *Practical Radiation Protection in Healthcare* ed C J Martin and D G Sutton (Oxford: Oxford University Press) ch 16.

McVey S, Sandison A and Sutton D G 2013 An assessment of lead eyewear in interventional radiology *J. Radiol. Prot.* **33** 647–59

Mehlmann C T and DiPasquale T G 1997 Radiation exposure to the orthopaedic surgical team during fluoroscopy: "how far away is far enough?" *J. Orthop. Trauma* **11** 392–8

Mesbahi A and Rouhani A 2008 A study on the radiation dose of the orthopaedic surgeon and staff from a mini C-arm fluoroscopy unit *Radiat. Prot. Dosimetry* **132** 98–101

Miller D L, Vañó E, Bartal G, Balter S, Dixon R, Padovani R, Schueler B, Cardella J F and de Baère T 2010 Occupational radiation protection in interventional radiology: a joint guideline of the Cardiovascular and Interventional Radiology Society of Europe and the Society of Interventional adiology *Cardiovasc. Interv. Radiol.* **33** 230–9

Müller L P, Suffner J, Wenda K, Mohr W and Rudig L 1996 Radiation burden to the hands of surgeons in intramedullary nailing *Unfallchirurgie* **22** 253–9

Müller L P, Suffner J, Wenda K, Mohr W and Rommens P M 1998 Radiation exposure to the hands and the thyroid of the surgeon during intramedullary nailing *Injury* **29** 461–8

Naidu L S, Singhai S, Preece D E, Vohrah A and Loft D E 2005 Radiation exposure to personnel performing endoscopic retrograde cholangiopancreatography *Postgrad. Med. J.* **61** 660–2

Neeman Z, Dromi S A, Sarin S and Wood B J 2006 CT fluoroscopy shielding: decreases in scattered radiation for the patient and operator *J. Vasc. Interv. Radiol.* **17** 1999–2004

Nickoloff E L, Khandji A and Dutta A 2000 Radiaition doses during CT fluoroscopy *Health Phys.* **79** 675–81

Noordeen M H, Shergill N, Twyman R S, Cobb J P and Briggs T 1993 Hazard of ionizing radiation to trauma surgeons: reducing the risk *Injury* **24** 562–4

NRPB 2001 *Guidance Notes for* Dental Practitioners *on the* Safe Use *of X-ray* Equipment (Didcot, UK: NRPB) https://www.gov.uk/government/uploads/system/uploads/attachment_data/file/337178/misc_pub_DentalGuidanceNotes.pdf (accessed May 2017)

O'Connor U, Gallagher A, Malone L and O'Reilly G 2013 Occupational radiation dose to eyes from endoscopic retrograde cholangiopancreatography procedures in light of the revised eye lens dose limit from the International Commission on Radiological Protection *Br. J. Radiol.* **86** 20120289

Olgar T, Bor D, Berkmen G and Yazar T 2009 Patient and staff doses for complex X-ray examinations *J. Radiol. Prot.* **29** 393–407

Padovani R, Le Heron J, Cruz-Suarez R, Duran A, Lefaure C, Miller D L, Sim H K, Vano E, Rehani M and Czarwinski R 2011 International project on individual monitoring and radiation exposure levels in interventional cardiology *Radiat. Prot. Dosimetry* **144** 437–41

Palácio E P, Ribeiro A A, Gavassi B M, Di Stasi G G, Galbiatt J A, Junior A D and Mizobuchi R R 2013 Exposure to the surgical team to ionizing radiation during orthopaedic surgical procedures *Rev. Bras. Ortop.* **49** 227–32

Rampersaud Y R, Foley K T, Shen A C, Williams S and Solomito M 2000 Radiation Exposure to the spine surgeon during fluoroscopy assisted pedicle screw insertion *Spine* **25** 2637–45

Ritter M, Krombach P, Martinschek A, Siegel F P, Schmitt M, Weiss C, Häcker A and Pelzer A E 2012 Radiation exposure during endourologic procedures using over-the-table fluoroscopy sources *J. Endourol.* **26** 47–51

Romanova K, Vassileva J and Alyakov M 2015 Radiation exposure to the eye lens of orthopaedic surgeons during various orthopaedic procedures *Radiat. Prot. Dosimetry* **165** 310–3

Van Rooijen B D, de Hann M W, Das M, Arnoldussen C W K P, de Graaf R, van Zwam W H, Backes W H and Jeukens C R L P N 2014 Efficacy of radiation safety glasses in interventional radiology *Cardiovasc. Interv. Radiol.* **37** 1149–55

Safak M, Olgar T, Bor D, Berkmen G and Gogus C 2009 Radiation doses of patients and urologists during percutaneous nephrolithotomy *J. Radiol. Prot.* **29** 409–15

Saidatul A, Azlan C, Megat A M, Abdullah B and Ng K 2010 A survey of radiation dose to patients and operators during radiofrequency ablation using computed tomography *Biomed. Imaging Interv. J.* **6** e1

Sandblom V, Mai T, Almén A, Rystedt H, Cederblad A, Båth M and Lundh C 2013 Evaluation of the impact of a system for real-time visualisation of occupational radiation dose rate during fluoroscopically guided procedures *J. Radiol. Prot.* **33** 693–702

Sanders R, Koval K J, DiPasquale T, Schmelling G, Stenzler S and Ross E 1993 Exposure to the orthopaedic surgeon to radiation *J. Bone Joint Surg.* **75** 326–30

Sauren L D, van Garsse L, van Ommen V and Kemerink G J 2011 Occupational radiation dose during transcatheter aortic valve implantation *Cathet. Cardiovasc. Interv.* **78** 770–6

Shortt C P, Al-Hashimi H, Malon L and Lee M J 2007 Staff radiation doses to the lower extremties in interventional radiology *Cardiovasc. Interv. Radiol.* **30** 1206–9

Singer G 2005 Occupational radiation exposure to the surgeon *J. Am. Orthop. Surg.* **13** 69–76

Sutton D G, Martin C J, Williams J R and Peet D 2012 *Radiation Shielding for Diagnostic Radiology* (London: British Institute of Radiology)

Sánchez R, Vano E, Fernández J M and Gallego J J 2010 Staff radiation doses in a real-time display inside the angiography room *Cardiovasc. Interv. Radiol.* **33** 1210–4

Sánchez R M, Vañó E, Fernández J M, Pifarré X, Ordiales J M, Rovira J J, Carrera F, Goicolea J and Fernández-Ortiz A 2016 Occupational eye lens doses in interventional cardiology. A multicentric study *J. Radiol. Prot.* **35** 133–43

Sung K H, Min E, Chung C Y, Jo B C, Park M S and Lee K 2016 Measurements of surgeons' exposure to ionizing radiation dose: comparison of conventional and mini C-arm fluoroscopy *J. Hand Surg. Eur.* **41** 340–5

Tasbas B A, Yagmurlu M F, Bayrakci K, Ucaner A and Heybeli M 2003 Which one is at risk in intraoperative fluoroscopy? Assistant surgeon or orthopaedic surgeon? *Arch. Orthop. Trauma Surg.* **123** 242–4

Taylor E R, Kramer B, Frye T P, Wang S, Schwartz B F and Köhler T S 2013 Ocular radiation exposure in modern urological practice *J. Urol.* **190** 139–43

Teeuwisse W M, Geleijns J, Broerse J J, Obermann W R and van Persijn van Meerten E L 2001 Patient and staff dose during CT guided biopsy, drainage and coagulation *Br. J. Radiol.* **74** 720–6

Tsalafoutas I A, Tsapaki V, Kaliakmanis A, Pneumaticos S, Tsoronis F, Koulentianos E D and Papachristou G 2008 Estimation of radiation doses to patients and surgeons from various fluoroscopically guided orthopaedic surgeries *Radiat. Prot. Dosimetry* **128** 112–9

Tsapaki V, Paraskeva K D, Mathou N, Andrikopoulos E, Tentas P, Triantopoulou C and Karagiannis J A 2011 Patient and endoscopists radiation doses during ERCP procedures *Radiat. Prot. Dosimetry* **147** 111–3

Vanhavere F *et al* 2012 ORAMED: Optimisation of Radiation Protection for Medical Staff *7th EURADOS Report* 2012-02 ISSN 2226-8057, ISBN 978-3-943701-01-2. Available at: http://www.eurados.org./en/Documents_Publications (accessed May 2017)

Vañó E, Fernandez J M, Resel L E, Moreno J and Sanchez R M 2016 Staff lens doses in interventional urology. A comparison with interventional radiology, cardiology and vascular surgery values *J. Radiol. Prot.* **36** 37–48

Vañó E, Fernandez J M and Sánchez R 2011 Occupational dosimetry in real time. Benefits for interventional radiology *Radiat. Meas.* **46** 1262–5

Vañó E, Fernández J M, Sánchez R M and Dauer L T 2013 Realistic approach to estimate lens doses and cataract radiation risk in cardiology when personal dosimeters have not been regularly used *Health Phys.* **105** 330–9

Whitby M and Martin C J 2003 Radiation doses to the legs of radiologists performing interventional procedures: are they a cause for concern? *Br. J. Radiol.* **76** 321–7

Whitby M and Martin C J 2005 A study of the distribution of dose across the hands of interventional radiologists and cardiologists *Br. J. Radiol.* **78** 219–29

Worrall M, McVey S and Sutton D G 2012 Proposals for revised factors in intra-oral radiography shielding calculations *Br. J. Radiol.* **32** 243–9

Zagorska A, Romanova K, Hristova-Popova J, Vassileva J and Katzarov K 2015 Eye Lens exposure to medical staff during endoscopic retrograde cholangiopancreatography *Phys. Med.* **31** 781–4

Colin John Martin, David Temperton, Anthony Hughes and Thomas Jupp

Chapter 6

Radiotherapy

6.1 Introduction

There is a wide variety of techniques used for radiotherapy treatment and this chapter considers the likely doses and requirements for personal monitoring with each. Equipment will generally be operated in dedicated facilities and a number of techniques and systems will be in place to ensure that staff do not enter the facility when radiation is being emitted. In these situations, although staff will not receive measureable doses during routine work, there may be risks of exposure resulting from an accident or incident, for which personal monitoring arrangements may need to be in place.

6.2 External beam radiotherapy

Radiation protection in any well-designed radiotherapy facility will be primarily based on remote working. Engineering controls such as interlocks prevent inadvertent access to the room whilst the radiation is on. Staff should always be outside the room when the patient is being treated and the facility will have been designed with the use of appropriate room shielding to limit radiation exposure to staff outside the treatment room. This will of course be augmented by systems of work and local rules to ensure that only trained persons work in the area, supervising anyone else who is not trained. In such instances, occupational doses are restricted to very low levels by the careful consideration of shielding calculations in the planning stage, so personnel monitoring over time should yield results which will not be measurable. Even so, there is still a role for personal monitoring in external beam radiotherapy.

doi:10.1088/2053-2563/aaddb6ch6

It is recommended that staff operating, maintaining, and testing external beam therapy equipment are monitored with a body dosemeter to detect any inadvertent exposure from situations such as those described in the following paragraphs.

Dose monitoring of individuals is required for the following situations;

(a) To record inadvertent exposures caused by a breakdown in engineering controls and/or systems of work which could have significant consequences in radiotherapy.

(b) To assess exposures from induced radioactivity within the treatment room for linacs operating at energies >10 MV. Petrovic *et al* (2011) found significant traces of ^{28}Al, ^{24}Na, ^{56}Mn, ^{54}Mn, ^{187}W, ^{64}Cu and ^{62}Cu in a treatment room following irradiation with an 18 MV beam. The additional annual dose to staff ranged from 0.25 to 1.7 mSv and was strongly influenced by delaying entrance to the therapy room by 2–4 min, as the measured dose rate decreases rapidly as a function of time after the beam has been turned off.

(c) Assessment of occupational dose incurred by therapy radiographers during PET/CT treatment planning. The external dose rate around a patient administered with 300 MBq ^{18}FDG one hour post injection can range from 15–35 μSv h^{-1} at a distance of 0.5 m (Leide-Svegborn 2010). Given the relatively long time required for accurate positioning and marking of the patient there is potential for attendant staff to receive whole body doses of 4–6 μSv per patient, which could translate into annual doses of $\geqslant$1 mSv for workloads of 4 patients per week. Passive whole body dosemeters worn at the waist are equally reliable for monitoring γ doses from PET tracers. In this situation occupational doses to the eyes and hands are unlikely to be very different from the whole body dose, so additional extremity monitoring is probably not justified.

Personal monitoring will highlight changes in dose levels around facilities where there has been a change of equipment or usage. For instance, increased monitor unit delivery associated with intensity modulated radiotherapy (IMRT) is accompanied with elevated levels of scattered and leakage radiation, possibly leading to increases in occupational exposure (Horton and Eaton 2017). Mutic *et al* (2001) has shown that IMRT requires an increase in secondary barrier shielding. Additionally, the introduction of flattening filter-free linacs into existing bunkers needs to be evaluated, even though there is evidence that instantaneous dose rates outside the treatment room can be substantially reduced with a flattening filter-free accelerator (Vassiliev *et al* 2007, Jank *et al* 2014).

Radioactivity may be induced in the treatment head of linear accelerators operated at over 10 MV. Dose rates will decline within minutes, but the dose rates in areas that would be occupied by the hands of engineers during maintenance should be measured and results incorporated into risk assessments prior to procedures being agreed. Finger monitoring should be undertaken to check doses to the finger tip, if these are likely to exceed 2–4 mSv per month.

6.3 HDR/PDR brachytherapy

Most modern brachytherapy treatments are delivered by remote high dose rate (HDR) or pulsed dose rate (PDR) afterloading systems, where staff are not present in the treatment room while the sources are exposed. In such situations personal monitoring of staff will not record measurable doses provided the unit has been appropriately shielded. However, there may still be occasions where monitoring will be required.

Any scenario requiring staff intervention where the treatment source has not returned to its shielding will demand personal monitoring. Such situations could occur during QC procedures, source exchange or during the treatment of a patient. Whenever a HDR source is unshielded the external dose rate can be very high. For a 370 GBq ^{192}Ir HDR source the expected dose rate at 1 m can reach levels of 0.7 mGy min^{-1} (Horton *et al* 2015). The risk assessment may demonstrate the need for classification to deal with such situations.

> For personal dose monitoring of staff who may find themselves in situations where they need to deal with an unshielded brachytherapy source, an instant read-out EPD is recommended for recording whole body dose. The EPD needs to have a measurement range of at least 0.01–10 mSv and a suitable response to the γ-radiation emitted by the source. This must be worn in addition to a passive whole body dosemeter.

An advantage of an EPD is that it may be possible for an audible alarm to be activated at a particular dose rate which, if appropriately chosen, may assist staff in minimising their dose during source recovery.

> Staff routinely involved in the transportation of HDR sources will require whole body monitoring to ensure adequate management of doses resulting from the exposure to external radiation. For a ^{192}Ir HDR source the dose rate at 1 m can reach 10 μSv h^{-1}. A passive body dosemeter would suffice in these instances.

Additionally, there will be a requirement for periodic assessment of contamination of the packages and vehicles involved in the transportation.

6.4 Manual brachytherapy (temporary implants)

Manually afterloaded brachytherapy using ^{192}Ir wire may still be in use in some centres in the UK. Pre-defined lengths of ^{192}Ir wire with a typical linear activity of

1–10 MBq mm^{-1} are cut from a 50 cm long coil and inserted into plastic tubing for protection and support. Such procedures may typically involve implantation of 1-2 GBq of ^{192}Ir wire leading to external dose rates of 0.1–0.2 mSv h^{-1} at 1 m from the site of the implant (Bomford and Kunkler 2003, Bownes 2011). Additionally, the dose rate close to the wire is likely to be in the range of 10–20 mGy h^{-1} for a 1 cm length containing 50 MBq (Cheung *et al* 1997). Hence, routine monitoring of extremity doses is advisable. The optimum location of a finger dosemeter for this type of work has not been established, so the position of dosemeters should be determined through observation. Contamination monitoring of all work areas with a scintillation probe will also be required once source preparation has been completed to detect any fragments of wire and prevent potential inadvertent skin exposures.

Temporary ^{192}Ir implants are typically *in situ* for 3–5 days and consequently, nursing staff will need to restrict their contact times with the patient during this period. Because of the relatively large external dose rates whole body monitoring of nursing staff will be required for reassurance purposes and to confirm that systems of work are being followed. Rather than issue multiple dosemeters to a number of nurses, a single EPD may be more suitable here as it will provide an instant dose reading that can be recorded by each wearer to create individual dose records.

> For manual afterloading brachytherapy, monitoring with a dosemeter on the chest or collar that will give an indication of body and eye dose is recommended for those preparing the sources and for nursing staff attending the patient. Finger dose monitoring is also recommended for those preparing the sources.

6.5 Permanent iodine seed implants

Permanent implantation of ^{125}I seeds has become an increasingly attractive treatment option for cancer of the prostate. In this procedure ^{125}I seeds are implanted using an applicator under fluoroscopic and transrectal ultrasound guidance to visualise the prostate (Yoshioka 2009). In any one patient, 40–100 seeds each containing 9–15 MBq ^{125}I may be implanted at a time. During the implantation procedure surgical staff will be exposed to the low energy γ-radiation emitted by ^{125}I (average energy = 28 keV) as well as scattered radiation from the fluoroscopy unit. Whole body doses to theatre staff (measured using TLDs on the thorax and abdomen under a 0.25 mm Pb equivalent apron) in the range 10–30 μSv per implantation have been reported (Anglesio *et al* 2005, Fujii *et al* 2011). Fujii *et al* (2011) also measured the dose to arms and hands of the implanting surgeon, wearing lead gloves, of 0.03 mm Pb nominal thickness. They found a maximal hand dose per implantation procedure of 0.65 mSv recorded on the outside of the glove; doses inside the glove were 50%–70% lower. Other studies have returned mean doses of 1–2 mSv to the fingertips and 0.6–1 mSv to the back of the dominant hand of surgeons per operation (Schiefer *et al* 2009).

Extremity monitoring of surgeons is therefore recommended for this procedure. The position of the maximum dose on the hands has yet to be determined and therefore the location for monitoring should be decided through observation. Monitoring appears to be especially important when the procedure is first introduced into a centre as dose reductions of up to 50% have been observed during the first year of use (Schwartz *et al* 2003, Anglesio *et al* 2005, Fujii *et al* 2011).

> It is recommended that surgeons wear a body dosemeter, and finger stall dosemeters on the index fingers and thumbs of each hand when first carrying out this procedure. This should be repeated until a reliable dose assessment has been obtained, after which a regime for routine monitoring should be decided. If monitoring is necessary, it may be performed with a ring dosemeter, but ratios for the tip to ring doses may need to be established.

Effective doses to other theatre staff have been reported as <10 μSv per patient through appropriate use of shielded screens and maintenance of large distances from the patient (Fujii *et al* 2011).

> Theatre staff assisting with iodine seed implants do not require dose monitoring.

Technical staff preparing seeds for implantation are also likely to be exposed, but with appropriately designed and shielded work areas it is finger dose that is likely to be of most concern. van Haaren *et al* (2011) measured the dose on the index finger nails of both hands for 30 workers preparing ^{125}I seed implants in three different hospitals. The measured doses per procedure varied from <30 μSv at one hospital to an average of 300 μSv at the other two. In this study doses were corrected for self-attenuation of radiation in the finger by multiplying the measured results by a factor of 2. Anglesio *et al* (2005) also measured finger dose incurred during source preparation as 65 μSv per procedure, although the exact location of the dosemeter was not mentioned.

> Monitoring of finger doses is recommended for staff preparing iodine seeds for implantation.

After implantation patients are often returned to a ward for up to 48 hours prior to discharge. During this time nursing staff will be exposed while caring for the patient. Measured dose rates of 40–55 μSv h^{-1} on the anterior skin surface have been reported, but these levels drop rapidly with increasing distance to 5–6 μSv h^{-1} at 30 cm. Dose rates at the lateral surface of the pelvis are very low; ~2.5 μSv h^{-1} due to self-attenuation with the body (Anglesio *et al* 2005). Nursing staff coming into close contact with the patient for 1 hour per day could therefore expect to receive an effective dose of ~10 μSv per patient over a 48 h period. This would translate into an annual dose of 0.5 mSv for a workload of 50 patients per year, however, this dose could be greatly reduced by providing ward staff with a lead apron. In this instance, reassurance monitoring would be recommended in the early stages of an ^{125}I implantation service. An EPD worn at the collar of the lead apron would be suitable and provides a flexible choice of dosemeter for monitoring the dose to the body and eyes of a number of different nurses.

6.6 Ophthalmic applicators

Treatment of uveal melanoma and retinoblastoma using ^{106}Ru or ^{90}Sr eye plaques has been practiced for over 20 years. Applicators are available in a number of different sizes to provide a match to the individual tumour size and location. They typically have the shape of a segment of a sphere with similar dimension to the eye ball, with the radius of the segment being 12–14 mm, and have special eyelets to be sutured to the sclera. These devices emit β-radiation only with a maximum energy of 3.5 MeV (^{106}Ru) and 2.3 MeV (^{90}Sr) and are shielded by 17 mm and 10 mm of Perspex respectively (Delacroix *et al* 2002). Monitoring of occupational whole body dose is not appropriate given the depth of penetration of the radiation but significant doses to the skin or eye could be incurred if not handled carefully.

During routine operation the plaque is manipulated using a holder which consists of a Perspex shielding plate, a fixation ring for the shielding plate and a holder with a mechanism for connecting to the source. The shielding plate is sufficiently thick to absorb any β-radiation emitted in the direction of the operator and consequently the operator skin dose should not be measureable. However, the equivalent dose rate near the active (treatment) surface of the plaque can reach levels of 20–40 mGy s^{-1} (Cohen *et al* 2013) so extreme caution is required when handling a plaque outside of its shielding. The sources will require frequent wipe testing and visual inspection for damage to the treatment surface. Wipe tests will need to be performed using long-handled forceps because of the very high local dose rates, while visual inspection of the active surface can only be done through a Perspex plate thick enough to absorb all the β-ray-radiation.

There is a lack of published data on actual doses received by staff, but given the potential for high doses it would be prudent to audit finger doses to the operator during therapy and also any technical staff assessing the sources for damage/leakage. The extremity monitors must have a suitable beta response. Maximum doses are likely to be received by the fingertips of the hand used to hold the applicator so monitoring should involve finger stalls. The intervals for audit will be determined by the workload and changes in theatre staff involved in the procedure.

Initial monitoring of the dose to the fingertip using finger stalls is recommended for staff handling and inserting eye plaques to establish whether regular monitoring is required.

6.7 Intraoperative radiotherapy

Intraoperative radiotherapy (IORT) involves direct irradiation of the tumour bed during surgery and has been in clinical use for over 30 years (Calvo *et al* 2000, Beddar *et al* 2006). More recently there has been interest in the use of IORT as a technique of partial breast irradiation for women with early breast cancer (Hanna and Kirby 2015) and given the availability of newer and lower cost treatment devices there could be increased integration of this technique into clinical practice. IORT may be delivered using a range of physical techniques (Gunderson *et al* 2011) including:

- Intraoperative electron beam irradiation using mobile linear accelerators,
- Intraoperative brachytherapy using ^{192}Ir HDR units, and
- Electronic brachytherapy using low kilovoltage x-rays machines.

Each of these techniques will present specific radiation protection challenges, however, in the UK IORT has been delivered solely using mobile kilovoltage x-ray machines (Eaton *et al* 2013). These devices emit x-rays with a mean energy of ~30 keV and irradiate the tumour bed in a single fraction using different sized applicators. Treatment times are typically 20–25 min in duration and require radiotherapy and anaesthetic personnel to remain in the theatre (usually behind a shielded screen) to monitor both treatment delivery and patient response. The x-ray energy is similar to that used in diagnostic procedures and the output dose rate when averaged over the treatment time is typically ~10 mGy h^{-1} at 1 m from the unshielded source (Eaton *et al* 2011). In spite of these high dose rates, the soft x-rays are easily shielded and published data from radiation surveys conducted during IORT suggest that occupational whole body doses are typically 1–2 μSv per procedure (Eaton *et al* 2011, Mobit *et al* 2015).

> Systems of work which are in place to ensure that staff remain behind the protective screen should be appropriately supervised, and whole body monitoring will provide reassurance that the measures are effective. Either an EPD or a passive dosemeter worn at the collar would suffice for this purpose and also allow an assessment of eye dose, although the response of any detector to the low energy photons must be established beforehand. There is no indication that extremity monitoring is required for theatre staff in these procedures.

References

Anglesio S, Calamia E, Fiandra C, Giglioli F R, Ragona R, Ricardi U and Ropolo R 2005 Prostate brachytherapy with iodine-125 seeds: radiation protection issues *Tumori* **91** 335–8

Beddar A S, Biggs P J, Chang S, Ezzell G A, Faddegon B A, Hensley F W and Mill M D 2006 Intraoperative radiation therapy using mobile electron linear accelerators: Report of AAPM radiation therapy committee task group No. 72 *Med. Phys.* **33** 1476–89

Bomford C K and Kunkler I H 2003 *Walter & Millers Textbook of Radiotherapy* 6th edn (Edinburgh: Churchill Livingstone)

Bownes P 2011 Radiation protection issues in brachytherapy *Radiotherapy in Practice—Bracytherapy* ed P J Hoskin (Oxford: Oxford University Press)

Calvo F A, Hoekstra H J and Lehnert T 2000 Intraoperative radiotherapy: 20 years of clinical experience, technological development and consolidation of results *Eur. J. Surg. Oncol.* **26** S1–4

Cheung Y C, Yu P K N, Young E C M, Chan C L, Ng M F and Tang F 1997 The dose distribution close to an 192Ir wire source: EGS4 Monte Carlo calculations *Phys. Med. Biol.* **42** 401–6

Cohen V M L, Papastefanou V P, Liu S, Stoker I and Hungerford J L 2013 The use of Strontium-90 beta radiotherapy as adjuvant treatment for conjunctival melanoma *J. Oncol.* **2013** Article ID 349162

Delacroix D, Guerre J P, Leblanc P and Hickman C 2002 Radionuclide and radiation protection data handbook 2nd edition *Radiat. Prot. Dosimetry* **98** 1–168

Eaton D J, Earner B, Faulkner P and Dancer N 2013 A national dosimetry audit of intraoperative radiotherapy *Br. J. Radiol.* **86** 20130447

Eaton D J, Gonzalez R, Duck S and Keshtgar M 2011 Radiation protection for an intra-operative X-ray device *Br. J. Radiol.* **84** 1034–9

Fujii K, Ko S, Nako Y, Tonari A, Nishizawa K, Akahane K and Takayama M 2011 Dose measurement for medical staff with glass dosemeters and thermoluminescent dosemeters during ^{125}I brachytherapy for prostate cancer *Radiat. Prot. Dosimetry* **144** 459–63

Gunderson L L, Willet C G, Calvo F A and Harrison L B 2011 *Intraoperative Irradiation: Techniques and Results* 2nd edn (New York: Humana)

Hanna G G and Kirby A M 2015 Intraoperative radiotherapy in early stage breast cancer: potential indications and evidence to date *Br. J. Radiol.* **88** 20140686

Horton P and Eaton D 2017 *Design and Shielding of Radiotherapy Treatment Facilities* (York: IPEM)

Horton P, Peet D J, Sutton D G and Martin C J 2015 Radiotherapy: brachytherapy and unsealed radionuclide therapy *Practical Radiation Protection in Healthcare* ed C J Martin and D G Sutton (Oxford: Oxford University Press) ch 20

Jank J, Kragl G and Georg D 2014 Impact of a flattening filter free linear accelerator on structural shielding design *Z. Med. Phys.* **24** 38–48

Leide-Svegborn S 2010 Radiation exposure of patients and personnel from a PET/CT procedure with ^{18}FDG *Radiat. Prot. Dosimetry* **139** 208–13

Mobit P N, Rajaguru P, Brewer M, Baird M, Packianathan S and Yang C C 2015 Radiation safety consideration during intraoperative radiation therapy *Radiat. Prot. Dosimetry* **164** 376–82

Mutic S, Low D A, Klein E E, Dempsey J F and Purdy J A 2001 Room shielding for intensity-modulated radiation therapy treatment facilities *Int. J. Radiat. Oncol. Biol. Phys.* **50** 239–46

Petrovic N, Krestic-Vesovic J, Stojanovic D, Ciraj-Bjelac O, Lazarevic D and Kovacevic M 2011 Contribution of activation products to occupational exposure following treatment using high-energy photons in radiotherapy *Radiat. Prot. Dosimetry* **143** 109–12

Schiefer H *et al* 2009 Exposure of treating physician to radiation during prostate brachytherapy using iodine-125 seeds *Strahlenther. Onkol.* **185** 689–95

Schwartz D J, Davis B J, Vetter R J, Pisansky T M, Herman M G, Wilson T M, LaJoie W N and Oberg A L 2003 Radiation exposure to operating room personnel during transperineal interstitial permanent prostate brachytherapy *Brachy* **2** 98–102

van Haaren P M A, van't Riet A, Moerland M A, Koedooder C and Westendorp H 2011 Dose to fingertips of staff preparing stranded iodine-125 seeds for permanent prostate implants *Radiat. Prot. Dosimetry* **145** 61–5

Vassiliev O N, Titt U, Kry S F, Mohan R and Gillin M T 2007 Radiation safety survey on a flattening filter-free medical accelerator *Radiat. Prot. Dosimetry* **124** 187–90

Yoshioka Y 2009 Current status and perspectives of brachytherapy for prostate cancer *Int. J. Clin. Oncol.* **14** 31–6

IOP Publishing

Guidance on the Personal Monitoring Requirements for Personnel Working in Healthcare

Colin John Martin, David Temperton, Anthony Hughes and Thomas Jupp

Chapter 7

Risk assessments to predict likely personal doses

7.1 Introduction

Dose monitoring practices will normally be based on recommendations by the radiation protection adviser. The level of dose monitoring required in areas in which doses are variable should be evaluated through risk assessment. This chapter provides guidance on good practice to fulfil requirements of the Ionising Radiations Regulations (2017). The monitoring performed should depend on a combination of the dose level, the potential to receive higher doses and the likely variation in doses recorded. If someone could receive a high dose at any time from a technically demanding interventional procedure, or a difficult radiopharmaceutical injection, then it may be appropriate to monitor their eyes and/or hands all the time.

The main recommendations about monitoring different parts of the body are summarized in table 7.1, together with the dose limit and the level at which radiation workers must be classified, and these may be used for staff working in any area. More detailed specific recommendations on monitoring requirements for different applications are set out in table 3.1 for nuclear medicine and table 4.2 for diagnostic and interventional radiology. The information contained includes a combination of requirements for initial measurements to determine the levels of dose received and ones for regular monitoring which will depend to some extent on results from initial monitoring. For some staff who assist in a few interventional procedures or inject radiopharmaceuticals for a limited number of patients, regular monitoring may only be required using a body dosemeter worn at the collar or on the chest to give indications of body and eye doses, provided recommended precautions are observed. Whereas others who perform these tasks on a regular basis may need to wear dedicated eye and hand dosemeters as well.

An initial dose assessment may be called for in order to check dose levels. The only reliable way to obtain data on doses received by personnel is from personal

Table 7.1. Recommendations on approaches to dose monitoring at different dose levels (based on information reported in Martin and Whitby 2003, Martin and Sutton 2015).

Tissue and dose quantity	Dose level for classification	Assessed Annual dose (mSv)	Assessed monthly dose (mSv)	Dose monitoring strategy
Body $H_p(10)$	6	$\geqslant 1$	$\geqslant 0.1$	Monitoring at chest, waist or collar depending on exposure situation
Eyes $H_p(3)$	15	1–6	0.2–0.5	Initial monitoring with dosemeter worn at collar or on the chest above any protection
Eyes $H_p(3)$		6–10	0.5–1	Regular monitoring with collar or head dosemeter is recommended
Eyes $H_p(3)$		$\geqslant 10$	$\geqslant 1$	Regular monitoring with head dosemeter is recommended
Extremities $H_p(0.07)$	150	20–50	2–4	Assessment of most exposed area and initial monitoring to establish dose levels
Extremities $H_p(0.07)$		> 50	> 4	Regular monitoring. Establish hand ratios to correct dose, if ring dosemeter does not measure most exposed area

monitoring, but results from passive dosemeters that are used for the majority of staff will only become available sometime after the exposure has occurred. An indication of potential dose is required initially, when practices are introduced. Various rules of thumb can be used to estimate likely dose levels for the purpose of carrying out a radiation risk assessment. Data are presented in tables 7.2–7.5 in the form of indicative doses per procedure, or based on the amount of radiation used, such as in the form of the kerma-area product (KAP) for radiology procedures or amount of radioactivity for nuclear medicine in order to assist the reader in dose estimation. The dose that someone receives will depend on the number of procedures which they perform or are involved in, but it will also be influenced by the position in which they stand with respect to patients and radiation sources, and the use of protective devices. Therefore, estimates of workload and some knowledge of the practical arrangements in performance of the procedures are required to inform the assessment. Once this has been obtained, initial monitoring requirements can be recommended using the doses derived, and requirements revised when personal dosimetry results become available. Even if monitoring results indicate that additional dosemeters do not have to be worn routinely, further monitoring may be required from time to time to ensure that techniques have not changed in a way that increases the dose levels. There are differences between the various specialties and these will be reviewed in the following sections. The aim of this chapter is to provide data on dose levels which might be used in calculations for use in risk assessments. Initial monitoring requirements can then be determined, and as more data are accumulated, strategies for dose monitoring and optimization developed. As more

Table 7.2. Indicators for deriving dose estimates for use in cardiology/radiology risk assessments (based on data reported in Vanhavere *et al* 2012, Martin and Magee 2013, Martin and Sutton 2015, Whitby and Martin 2003, 2005b, Martin 2009).

Tissue	Type of procedure or arterial access routes	Dose per procedure (μGy)	Doses/KAP (μGy (Gy cm^2)$^{-1}$)	KAP per month[a] for monitoring (Gy cm^2)
Eye	Cardiology: majority of procedures	40–60	1	500
Eye	Valvular and structural heart disease	100	1	500
Eye	Embolizations	180	3	200
Eye	Angiography of lower limb	60	5	100
Eye	Pacemaker insertions	60	6	100
Eye	ERCP	40	2	300
Eye	Other mobile C-arm applications	40–400	6	100
Hand	Percutaneous procedures		50	100
Hand	Radial access		10	500
Hand	Femoral access		5	1000
Hand	Angiography lower limb		40	100
Leg	Procedures without protective drape		10	500

[a]Workload above which dose monitoring should be considered.

Table 7.3. Indicative values for KAP and DLP for different types of procedure.

Fluoroscopy procedure	KAP (Gy cm^2)
Cardiac catheterisation procedure	30–50
Barium enema	20
Barium meal	12
ERCP	10
Radiograph of abdomen, pelvis or lumbar spine	2
CT examination	**DLP (mGy cm)**
Body	1000
Head	970

monitoring data become available, the values reported here can be improved upon and refined.

It should be borne in mind that dose monitoring is required not only for measurement of doses accrued gradually from routine practices, but also to record doses from any accidental exposure. Therefore, if a staff member is working in an

Table 7.4. Dose rates per unit activity from a point source and from unit activity administered for nuclear medicine procedures, at different distances from the anterior mid-trunk of adult patients (based on data reported in Smith and Stabin 2012, Mountford and O'Doherty 1999, Greaves and Tindale 1999, Benetar *et al* 2000, Greaves and Dunn 2014 and local measurements).

Radionuclide and scan phase[a]	Dose rates ($\mu Sv\ h^{-1}\ MBq^{-1}$)			
	1 m from point source	0.1 m from patient	0.5 m from patient	1.0 m from patient
^{18}F after injection	0.138	0.5	0.3	0.11
^{18}F uptake phase			0.25	0.09
^{18}F during scan			0.2	0.06
^{67}Ga	0.019	0.2	0.03	0.01
^{99m}Tc	0.019	0.3	0.05	0.02
^{111}In	0.081	1.0	0.2	0.06
^{201}Tl	0.011	0.2	0.07	0.02
^{123}I	0.042	0.6	0.1	0.03
^{131}I	0.052	1.5	0.2	0.06

[a]The values apply to the amount of activity at the time of injection, except for those for ^{18}F which relate to the average values during different phases of the examination process.

Table 7.5. Selection of values for doses to the fingertip from preparation and injection of imaging radionuclides.

Study	Radionuclide	Task	Median finger dose ($\mu Sv\ GBq^{-1}$)	Range of finger doses ($\mu Sv\ GBq^{-1}$)
Covens *et al* (2007)	^{99m}Tc	Preparation	65	20–110
Covens *et al* (2007)	^{99m}Tc	Administration	49	40–60
Carnicer *et al* (2011)	^{99m}Tc	Preparation	250	30–2060
Carnicer *et al* (2011)	^{99m}Tc	Administration	120	10–950
Martin (2016a) (no shield)	^{99m}Tc	Administration	35	17–230
Martin (2016a) (syringe shield)	^{99m}Tc	Administration	13	3–920
Covens *et al* (2007)	^{18}F	Preparation	570	290–850
Covens *et al* (2007)	^{18}F	Administration	320	200–750
Carnicer *et al* (2011)	^{18}F	Preparation	830	100–4430
Carnicer *et al* (2011)	^{18}F	Administration	640	140–4110

environment where a significant exposure could occur, and it would be difficult to quantify such an exposure, then dose monitoring should still be carried out, no matter how low the doses are that are received from routine work practices.

7.2 Risk assessment in interventional and fluoroscopy procedures

This section reviews the data available on dose levels for interventional procedures, where the operator is often adjacent to the patient and there may be other staff within the x-ray room while exposures are taking place. It considers how the data can be used in the prediction of the doses personnel are likely to receive.

7.2.1 Eye dose

Since staff involved with interventional and fluoroscopy procedures wear lead aprons to protect their body, the eye is the tissue that receives doses that might be anticipated to approach the classification level. There are large variations in the eye dose levels recorded for interventional clinicians. However, some indication of likely dose levels can be obtained from collation of results from multiple studies. Most record the dose and the number of procedures performed, therefore the simplest method of estimating dose would be through use of a dose per procedure. Many studies in recent years have also recorded the KAP for the purpose of patient dosimetry and this has been shown to be closely linked to the level of scattered radiation (Williams 1996, Sutton *et al* 2012). A link to the amount of radiation used should allow operator doses from a range of radiology procedures to be estimated. The mean or median values from a set of studies might at first sight be the best choice. However, since practices may not be optimized, dose ranges are large, so the third quartiles of distributions have been used to provide values for the initial assessments.

Cardiologists and radiologists performing interventional procedures are the staff closest to the radiation source and so the ones expected to receive the highest doses to the lens of the eye, so they will be considered first. Recommendations are based on a meta-analysis of 26 independent studies of interventional cardiology reported between 1996 and 2006 (Martin 2009, 2011) and a study by the European Radiation Group entitled ORAMED: Optimization of Radiation Protection of Medical Staff that collated dose monitoring data from investigations in nine European countries (Vanhavere *et al* 2012, Koukorava *et al* 2011, Martin and Magee 2013). The studies collated data on eye doses for a range of cardiology procedures and review of the third quartile doses suggests that a simple assumption of an average eye dose of 60 μSv per interventional cardiology procedure would be suitable for use in radiation risk assessments for more common procedures, including coronary angiography. However, a more conservative assumption is needed for certain procedures such as embolisations, angioplasties of the lower leg, and pacemaker insertions (Vanhavere *et al* 2012) and a summary of dose values for other procedures is given in table 7.2. Pacemaker insertions are also often performed using mobile C-arms in standard theatres with fewer shielding devices and less optimization of protection. Thus, although the patient dose levels may be lower, staff performing the procedures can receive high doses.

Dose levels for interventional procedures and other fluoroscopic procedures are variable, so it is more difficult to assign a dose per procedure to predict operator exposure. Therefore, it is helpful to consider assessments of eye doses based on KAP workload data where these are available. Data on operator eye dose linked to patient workload in terms of KAP from the literature have been collated (Martin and Magee 2013), including results from the ORAMED study (Vanhavere *et al* 2012). Based on the third quartile of the data distribution, an assumption that 1 Gy cm^2 equates to an eye dose of 1 μSv would be reasonable for the majority of procedures where protection is optimized. Results from the ORAMED study suggest that the relationship for certain procedures is sufficiently different, because of the position where the clinician stands with respect to the x-ray beam, to justify the use of larger factors. Suggested factors for use in initial risk assessments linking operator doses from scattered radiation to KAP workload are given in table 7.2 (Martin and Magee 2013). The higher dose values for pacemaker insertion and angiography of the lower leg are again linked to the use of mobile C-am units with less provision of specialised shielding devices. Similar arrangements also apply to theatre fluoroscopy using mobile C-arms which is often performed by clinicians other than radiology specialists, with more limited expertise. The uncertainty in these relationships is substantial and individual results could be at least four times higher if little attention is paid to radiation protection or four times lower if protective devices are fully utilized, so the factors given in table 7.2 can only provide indicative doses.

Results from other studies indicate that eye doses received by clinicians assisting in interventional procedures can vary from 30–100% of the dose received by the primary operator, while eye doses for nursing staff and radiographers are in the region of 20–40% and 5–30% respectively of that for the primary operator (Kicken *et al* 1999, Efstathopoulos *et al* 2003, Sanchez *et al* 2011, Antic *et al* 2013, Chida *et al* 2013, Omar *et al* 2017). However, this will be very dependent on the positions that they take up with respect to the patient couch and x-ray tube, and the location and use made of shielding devices.

Another area where an estimation of eye dose may be required is CT fluoroscopy. The dose received will be dependent on the position of the operator, since the gantry provides good protection in certain directions. It is difficult to estimate dose per procedure to staff in the scan room, as operator technique is a key factor. Dose levels within the main scatter field from the patient can be calculated from the dose length product (see section 7.2.3). Use of thyroid shields and lead glasses in addition to lead aprons is likely to be necessary for the operator. Careful monitoring of doses to eyes, hands, trunk and in some cases feet is recommended when a procedure is introduced to determine dose levels and ensure protection is optimized.

7.2.2 Extremity monitoring

Doses to the hands of clinicians performing interventional procedures vary with the route used to access the arterial system. Links have been established with the KAP workload, but these doses are variable as they depend strongly on the interventional operator's technique. Some factors are given in table 7.2 as a guide to whether monitoring is required to assist in preparing risk assessments. Training should be given in techniques and effective use of protective screens, protective patient drapes (e.g. Radpad, Lenexa, USA) and protective gloves in order to reduce hand doses, where these are found to be high. Some form of monitoring for the legs may be required occasionally where no protective screen is available, or where the operator's ankles and feet are not protected by the drape suspended from the couch (section 5.3.4).

7.2.3 Generalized method for prediction of dose levels from workload

Dose levels to which staff working with x-rays may be exposed for use in risk assessments can be estimated from their workload and their position with respect to the x-ray source in order to determine monitoring practice for the longer term. Scattered air kerma levels (K_S) at 1 m from the x-ray tube and patient can be calculated from the *KAP* using the equation:

$$K_s = S \times KAP \tag{7.1}$$

Where S is a scatter factor linked to the tube potential (Sutton *et al* 2012), which for scatter to the side from an undercouch PA exposure is given by the equation:

$$S_{\text{body}} = [(0.031 \times \text{kV}) + 2.5]\ \mu\text{Gy}(\text{Gy cm}^2)^{-1} \tag{7.2}$$

However, the scatter varies with angle and applying coefficients derived by Sutton *et al* (2012), scatter up towards the eye that is attenuated by passage through the patient's body would be given by:

$$S_{\text{head}} = [(0.014 \times \text{kV}) + 1.2]\ \mu\text{Gy}(\text{Gy cm}^2)^{-1} \tag{7.3}$$

Whereas scatter from the base of the couch towards the feet is substantially greater and will be given by the equation:

$$S_{\text{feet}} = [(0.05 \times \text{kV}) + 4]\ \mu\text{Gy}(\text{Gy cm}^2)^{-1} \tag{7.4}$$

The air kerma level (K) for unprotected areas of a member of staff at distance d from the patient i.e. the source of scatter will be given by:

$$K = K_s/d^2 \tag{7.5}$$

Approximate values of KAP for some common radiology examinations that can be used in assessing scatter doses for risk assessments are given in table 7.3. More detailed analyses of data for different procedures are given in Hart *et al* (2012) and Shrimpton *et al* (2014).

A similar approach can be used for calculating scatter air kerma levels for CT fluoroscopy from the dose length product (DLP) in regions outside the protection afforded by the scanner gantry (Wallace *et al* 2012). Here the scatter at 1 m from the iso-centre is given by:

$$K_S = S_{\text{CT}} \times \text{DLP} \tag{7.6}$$

Where S_{CT} is equal to 0.36 μGy (mGy cm)$^{-1}$ for body scans and 0.14 μGy (mGy cm)$^{-1}$ for head scans. The scatter air kerma at any position can then be calculated using equation (7.5).

7.3 Risk assessments for the use of radionuclides

7.3.1 Body doses arising from patients administered with radionuclides

Risk assessments for radionuclide users depend on the amount of activity that is handled and the level of protection that is in place. The time that staff members spend in close proximity to patients is significant for body exposure, and factors giving dose rates at different distances from patients taken from various studies are given in table 7.4. Dose rates will be lower when staff are further away and, although patients are not point sources, indicative dose rates at other distances can be derived by application of an inverse square law to the values at 1 m. These can be multiplied by estimations of the times and frequencies of exposure in order to derive values for doses that might be received. However, since times and distances are difficult to assess with any accuracy, such results will only give approximate dose levels. Point source dose rates are also given to enable calculations of exposure from partially shielded sources such as syringes and open vials. These values would normally be expected to be slightly higher than those from measurements on patients. The fact that this is not always the case probably reflects the accuracy with which measurements of dose rates from patients can be made in the hospital environment.

In order to evaluate the doses that individuals are likely to receive, the dose rate D_R for the relevant radionuclide from table 7.4 can be combined with the inverse square law for the distance d of the staff member from the patient and the time t for which they were in position to derive the dose from an exposure. The results for a series of exposures received (1 to i) can then be summed to calculate the body or eye dose received D_E over a period of time from an equation of the form:

$$D_E = \sum_i \frac{D_{Ri}}{d_i^2} \times t_i \tag{7.7}$$

The activity levels used in calculations should be adjusted for decay that will have occurred since injection. The calculated doses can then be used to provide a realistic starting point for risk assessments of eye or body dose.

7.3.2 Finger doses from preparation and injection of radionuclides

Vials and syringes should be shielded whenever possible. However, doses to parts of the hands can be high, when vials and syringes are handled directly. Moreover, there may be significant differences in dose levels in different parts of the hand, so the assessment of the most effective part to monitor should be part of the risk assessment. If the introduction of radioactivity can be delayed until more difficult parts of the procedure have been completed this will assist in keeping doses down. Placement of a venflon in a vein prior to introduction of radioactivity, can shorten the period of exposure to the active syringe, by removing the period when the staff member is attempting to insert the syringe needle into a viable vein from the exposure.

Any degree of accuracy in prediction of exposure levels is impossible without some measurements, as so much depends on technique. A selection of values for dose to the fingertip are given in table 7.5 from different studies to give an indication of the range. Results vary by more than a factor of ten. An initial conservative estimate for ^{99m}Tc radiopharmaceuticals might be 100 and 200 μSv GBq^{-1} for administration and preparation respectively, and 500 μSv GBq^{-1} for each stage with ^{18}F.

7.3.3 Radiopharmacies

In a radiopharmacy, the process of dispensing a radiopharmaceutical from a vial is repeated many times. The basic process involves withdrawal of radionuclide from the vial into a syringe and then injection of this into another vial. The regime adopted for sub-dispensing active vials can have a significant influence on the dose to the hand. A number of different options are possible. For example vials of activity for injection into individual patients may be dispensed directly from the eluate vial or first sub-dispensed into kit vials. The latter involves more manipulations and so more exposure from activity in the syringe, although there is less activity in the kit vials when they are dispensed. Since handling of activity in syringes tends to make up the biggest component of the dose, hand doses tend to be larger when more syringe manipulations are performed, particularly with higher activities (Whitby and Martin 2005a). However, the handling of vials containing larger activities and the potential for accidents also needs to be considered. Another option is to dispense radiopharmaceutical activity into a multi-vial that contains enough activity for several patient injections. A member of nuclear medicine staff is then required to withdraw appropriate volumes for injection into patients.

As already mentioned, the dose to the fingers during dispensing manipulations will be made up of components from the vial, from which activity is withdrawn, and the syringe into which activity is drawn. If during a dispensing manipulation the dose from a vial containing unit activity is D_v and the dose from a syringe containing unit activity is D_s, then the dose D_m from one manipulation involving the dispensing of activity A_k from an eluate vial containing activity $A_0 + A_k$ can be represented as:

$$D_m = D_v(A_0 + A_k) + D_s A_k \tag{7.8}$$

Then the withdrawal of a set of k nuclear medicine activities A_k from a vial initially containing activity $A_0 + kA_k$ can be represented by the dose D_k:

$$D_k = D_v(kA_0 + kA_k + (k-1)A_k + (k-2)A_k + \ldots\ldots + A_k) + kD_s A_k$$

$$= kD_v\left(A_0 + \frac{A_k(k+1)}{2}\right) + kD_s A_k \tag{7.9}$$

A similar equation can be applied to the dispensing of kit vials from eluate vials, or patient vials from kit vials (Whitby and Martin 2005a). It can therefore be used for comparisons of doses from different dispensing regimes to assess their likely impact on radiation doses to the fingers on radiopharmacy and nuclear medicine staff based on the relative activities handled.

The dose quantities D_v from the vial and D_s from the syringe will be substantially smaller if both the vial and syringe are effectively shielded.

7.4 Options for use in appraising personal monitoring results

The reliability of dose data from personnel monitoring can sometimes be questionable. Personal monitoring relies on individuals being conscientious in wearing dosemeters both consistently and in the correct locations. It is sometimes helpful to have additional information to confirm that personal monitoring results are correct.

7.4.1 Electronic dosemeters

Electronic dosemeters can be useful for monitoring body, eye and hand doses for limited time periods, and can be employed in evaluating protection measures during the early stages of the introduction of new procedures. They can provide data on doses received by staff within a short timescale and so can be useful in obtaining dosimetry data at an early stage during the introduction of a new technique that can contribute to the risk assessment. Their use in optimization is described in sections 3.5 and 5.5.2.

7.4.2 Calculations based on workload

One option in estimating dose levels for interventional procedures is to use factors linking known quantities such as KAP results for fluoroscopic procedures or amount of radioactivity handled to determine likely dose levels. Doses will depend on the positions where operators stand with respect to the x-ray tube or radionuclide source, and the distribution of scatter radiation. However, the dose values given in tables 7.2–7.4 for risk assessment purposes can also be compared with measured dose results to verify whether these are within the expected range, bearing in mind that they may be several times higher or lower. Dose rates from fluoroscopy procedures are higher where radiation is scattered from the surface of the patient, so when the tube gantry is angled, personnel standing adjacent to the x-ray tube will be exposed to higher levels of x-rays scattered from the patient. For CT fluoroscopy, the dose is very dependent on the position taken up with respect to the gantry, as this can provide effective protection, reducing the dose rate by a factor of ten (Wallace *et al* 2012). For nuclear medicine, doses received are strongly dependent on proximity to the patient. Inverse square law corrections can be used to derive approximate doses and dose rates at the distances where individuals stand and overall doses estimated using calculations based on equation (7.7) and similar.

7.4.3 Measurements using dosemeters attached to equipment

The calculation of dose rates depends on knowledge of how the x-ray equipment is used as well as the radiation emitted. Results are often required that will give information on the use of the x-ray equipment or radiation levels in the area. An option here is to use dosemeters in locations where they will be exposed to radiation fields similar to those that irradiate personnel. For example an active or passive dosemeter can be incorporated into a C-arm gantry to provide measures of scatter dose directly, which could be corrected to provide an assessment of likely doses to the eyes for interventional clinicians (Vañó *et al* 2013). This type of approach may be useful to confirm the validity of personal dosemeter results where routine wearing of dosemeters cannot be relied upon.

Vañó *et al* (2013, 2015) measured the dose with a dosemeter on the C-arm gantry at an angle of 45° from the x-ray tube, which is approximately 10 μGy $(\text{Gy cm}^2)^{-1}$. Since such a dosemeter will always be facing the patient the dose recorded will be substantially higher than that to which staff are exposed, but it will nevertheless give a reasonable indication of the level of radiation exposure from the x-ray equipment involved. Adjustments for the positions in which the operators stand should be based on an inverse square law, which provides a reasonable approximation to the actual decline in dose rate beyond a distance of 1 m. As a rough guide, multiplying the measured dose by a factor of 0.3 will give a worst case potential dose to the operator. However, if radiation protection is fully optimized, with use of ceiling suspended screens and other protective devices, the dose that the interventional operator receives should only be about one twentieth of the C-arm scatter dose (Vañó *et al* 2015). Thus ambient dosemeters attached to C-arms can help in the estimation of occupational doses, but they do not take account of the position adopted by the operator or their use of protective devices. They can only give an indication of whether doses recorded by personal dosemeters worn above the lead apron are realistic within a fairly broad range. Any more accurate personal dose assessment capable of replacing standard dosimetry techniques would also require technology for personnel position sensing.

Dosemeters can be placed at other locations within the x-ray room or the nuclear medicine department, if this is likely to provide a better assessment of the actual doses received by staff. However, if the dosemeter is further from the x-ray or scanning unit, it may be partially shielded by staff standing between the equipment and the dosemeter, and also the doses will be proportionately smaller so results will be less accurate. Another useful location for placing dosemeters can be in linac mazes and at maze entrances to measure neutron doses. Locations for dosemeters should be selected with care to ensure that any device to which they are attached provides a reasonable surrogate for staff positions and does not interfere with the procedures.

References

Antic V, Ciraj-Bjelac O, Rehani M, Aleksandric D, Arandjic D and Ostojic M 2013 Eye lens dosimetry in interventional cardiology: results of staff dose measurements and link to patient dose levels *Radiat. Prot. Dosimetry* **154** 276–84

Benetar N A, Cronin B F and O'Doherty M J 2000 Radiation dose rates from patients undergoing PET: impications for technologists and waiting areas *Eur. J. Nucl. Med.* **27** 583–9

Carnicer A *et al* 2011 Hand Exposure in diagnostic nuclear Medicine with 18F- and 99mTc-labelled radiopharmaceuticals—Results of the Oramed Project *Radiat. Meas.* **46** 1277–82

Chida K, Kaga Y, Haga Y, Kataoka N, Kumasaka E, Meguro T and Zuguchi M 2013 Occupational dose in interventional radiology procedures *Am. J. Roentgen.* **200** 138–41

Covens P, Berus D, Buls N, Clerinx P and Vanhavere F 2007 Personal dose m onitoring in hospitals: global assessment, critic al applications and future need *Radiat. Prot. Dosimetry* **124** 250–9

Efstathopoulos E P, Makrygiannis S S, Kottou S, Karvouni E, Giazitzoglou E, Korovesis S, Tzanalaridou E, Raptou P D and Katritsis D G 2003 Medical personnel and patient dosimetry during coronary angiography and intervention *Phys. Med. Biol.* **48** 3059–68

Greaves C D and Dunn M J 2014 The nuclear medicine patient *Practical Radiation Protection in Healthcare* ed C J Martin and D G Sutton (Oxford: Oxford University Press) ch 18

Greaves C D and Tindale W B 1999 Dose rate measurements from radiopharmaceuticals: implications for nuclear medicine staff and for children with radioactive parents *Nucl. Med. Commun.* **20** 179–87

Hart D, Hillier M C and Shrimpton P C 2012 *Doses to Patients from Radiographic and Fluoroscopic X-ray Imaging Procedures in the UK—2010 Review. HPA-CRCE-034* (Chilton: HPA)

Kicken P J H, Kemerink G J and van Engelshoven J M A 1999 Dosimetry of occupationally exposed persons in diagnostic and interventional arteriography. Part I: Assessment of entrance doses *Radiat. Prot. Dosimetry* **82** 93–103

Koukorava C, Carinou E, Simantirakis G, Vrachliotis T G, Archontakis E, Tierris C and Dimitriou P 2011 Doses to operators during interventional radiology procedures: Focus on eye lens and extremity dosimetry *Radiat. Prot. Dosimetry* **144** 482–6

Martin C J 2009 A review of radiology staff doses and dose monitoring requirements *Radiat. Prot. Dosimetry* **136** 140–57

Martin C J 2011 Personal dosimetry for interventional operators: when and how should monitoring be done? *Br. J. Radiol.* **84** 639–48

Martin C J 2016a Strategies for assessment of doses to the tips of the fingers in nuclear medicine *J. Radiol. Prot.* **36** 405–18

Martin C J and Magee J S 2013 Assessment of eye and body dose for interventional radiologists, cardiologists, and other interventional staff *J. Radiol. Prot.* **33** 445–60

Martin C J and Sutton D G 2015 Interventional radiology and cardiology *Practical Radiation Protection in Healthcare* ed C J Martin and D G Sutton (Oxford: Oxford University Press) ch 16

Martin C J and Whitby M 2003 Application of ALARP to extremity doses for hospital workers *J. Radiol. Prot.* **23** 405–21

Mountford P J and O'Doherty M J 1999 Exposure of critical groups to nuclear medicine patients *Appl. Radiat. Isot.* **50** 89–111

Omar A, Kadesjö N, Palmgren C, Marteinsdottir M, Segerdahl T and Fransson A 2017 Assessment of the occupational eye lens dose for clinical staff in interventional radiology, cardiology and neuroradiology *J. Radiol. Prot.* **37** 145–59

Shrimpton P C, Hillier M C, Meeson S and Golding S J 2014 *Doses from Computed Tomography (CT) Examinations in the UK – 2011 Review. HPA-CRCE-013* (Chilton: PHE)

Smith D S and Stabin M G 2012 Exposure rate constants and lead shielding values for over 1,100 radionuclides *Health Phys.* **102** 271–91

Sutton D G, Martin C J, Williams J R and Peet D 2012 *Radiation Shielding for Diagnostic Radiology* (London: British Institute of Radiology)

Sánchez R *et al* 2011 A national programme for patient and staff dose monitoring in interventional cardiology *Radiat. Prot. Dosimetry* **147** 57–61

The Ionising Radiations Regulations 2017 *SI 2017 No 1075, London HMSO* http://www.legislation.gov.uk/uksi/2017/1075/contents/made (accessed 9/2/18)

Vanhavere F *et al* 2012 ORAMED: Optimisation of Radiation Protection for Medical Staff *7th EURADOS Report* 2012-02 ISSN 2226-8057, ISBN 978-3-943701-01-2. Available at: http://www.eurados.org./en/Documents_Publications (accessed May 2017)

Vañó E, Fernández J M, Sánchez R M and Dauer L T 2013 Realistic approach to estimate lens doses and cataract radiation risk in cardiology when personal dosimeters have not been regularly used *Health Phys.* **105** 330–9

Vañó E, Sánchez R M and Fernandez J M 2015 Estimation of staff lens doses during interventional procedures. Comparing cardiology, neuroradiology and interventional radiology *Radiat. Prot. Dosimetry* **165** 279–83

Wallace H, Martin C J, Sutton D G, Peet D and Williams J R 2012 Establishment of scatter factors for use in shielding calculations and risk assessment for computed tomography facilities *J. Radiol. Prot.* **32** 39–50

Whitby M and Martin C J 2003 Radiation doses to the legs of radiologists performing interventional procedures: are they a cause for concern? *Br. J. Radiol.* **76** 321–7

Whitby M and Martin C J 2005a A multi-centre study of dispensing methods and hand doses in UK hospital radiopharmacies *Nucl. Med. Commun.* **26** 49–60

Whitby M and Martin C J 2005b A study of the distribution of dose across the hands of interventional radiologists and cardiologists *Br. J. Radiol.* **78** 219–29

Williams J R 1996 Scatter dose estimation based on dose-area product and the specification of radiation barriers *Br. J. Radiol.* **69** 1032–7

Colin John Martin, David Temperton, Anthony Hughes and Thomas Jupp

Chapter 8

Managing personal monitoring

8.1 General

The dosemeters used for personal monitoring should be carefully selected to ensure that they will adequately measure the radiation to which the employee will be exposed (see chapter 2). One of the biggest challenges associated with personal monitoring of staff in the healthcare sector is ensuring that staff reliably and regularly wear their dosemeters correctly and return the dosemeters for assessment at the end of the issue period. Recent professional body advice (Rogers *et al* 2017) written with support from HSE Specialist Inspectors (radiation) suggest that robust management systems for the use of personal monitors should include:

- written procedures describing how the system works including, who should be monitored, how the dosemeters are issued and collected back to ensure that they are returned on time, clearly defined responsibilities and performance indices (such as the number of late and missing dosemeters); and
- records to demonstrate that staff have been trained in the system, including how to wear their dosemeters and action to take when dosemeters are lost, accidentally exposed, etc;
- systems to track which dosemeters are returned by wearers and general compliance by staff;
- records of staff doses received;
- systems to ensure dosemeter results are promptly reviewed and suitable actions taken when investigation levels are breached.

The monitoring policy will need to include disciplinary action that will be taken to deal with non-compliance such as staff not wearing their dosemeters. Approved dosimetry services are able to provide information on dosemeter returns and

doi:10.1088/2053-2563/aaddb6ch8

recorded doses, often via on-line access to their records. They may also enable users to set specific local investigation levels.

8.2 Wearing and looking after the dosemeter

There is no legal requirement for the employer to provide a dosemeter for non-classified workers. However, where non-classified persons enter controlled areas, employers must put arrangements in place to demonstrate that any exposures received are ALARP and will not exceed the dose levels specified in regulation IRR17 Reg19(4), i.e. doses requiring classification or exceeding dose limits. In order to meet this requirement (and reassure staff) employers often decide to provide personal dose monitors. If employees are issued with a dosemeter, the law requires them to wear and look after the dosemeter and return it promptly for assessment after the end of the wear period (HSE 2001). Where an employer has arranged the issue of a personal monitor to an employee and as such requires the employee to use and wear the personal dosemeter, the employee must comply with those instructions under section 7 of the Health and Safety at Work Act 1974 and Reg. 35 of the IRR17. Employers should check that employees are complying with the required control measures and challenge and correct unsafe behaviour (HSE 2018).

Staff should be trained (IRR17 Reg 15) to know the importance of:

- looking after their dosemeter;
- informing the person who distributes the dosemeters as soon as practicable if
    - their dosemeter (or the holder) is damaged, accidentally exposed to radiation when it was not being worn, or has been exposed to excessive heat, moisture or chemicals,
    - they have lost or misplaced their dosemeter, or
    - they suspect that they have been exposed inadvertently (so the employer and ADS know the likely cause if an elevated reading is measured),
    - they find a dosemeter (e.g. in an x-ray room) that has presumably been lost.

They should give details of when they suspect that the loss or damage happened or where and when a dosemeter was found. The ADS and line manager should be advised accordingly.

Staff are more likely to take care of their dosemeters and return them at the correct times if they receive regular reports of the doses that are assessed by the dosimetry service. Although the majority of doses assessed by the dosimetry service are below recording level, it is important this information is still passed on to wearers, so they are aware that the dosemeters have been properly returned for assessment.

8.3 Missing dosemeters

Inevitably some dosemeters will be lost or damaged. For classified workers, the Approved Dosimetry Service (ADS) keeping the dose records (ADS Records) will require an estimated dose.

If it is not possible to accurately assess the dose received by the dosemeter because the dosemeter is lost, damaged, accidentally exposed, or for some other reason, IRR17 Regulation 23 stipulates that the employer must make an investigation (in conjunction with the Radiation Protection Adviser where appropriate), and either:

(a) estimate the dose and arrange for the ADS (Records) to enter the 'estimated dose' in the relevant dose record (NB the ADS must be supplied with a summary of the information used to estimate that dose); or
(b) exceptionally, where an estimated dose is not possible, arrange for the ADS (Records) to enter a 'notional dose' in the relevant dose record.

The ADS will request estimated doses whenever there is a gap in the dose records but will enter a notional dose if no reply is received. The notional dose equals the proportion of the annual dose limit relevant to the wear period—e.g. for a monthly whole body dosemeter, the notional dose would be 1.7 mSv.

For non-classified staff, there is no legal requirement to have dose records. However, it is common practice for the ADS to put some kind of average dose value into the person's dose record. For example, the average may be evaluated from the three previous dose assessments for that wearer. Even for non-classified wearers, the HSE would expect a robust management system to be in place to minimise the number of missing dosemeter results.

8.4 Application of dose correction factors

In certain circumstances a true assessment of the doses that a person receives is difficult to achieve with a dosemeter. In these circumstances it may be necessary to apply a correction factor to the dose measured on a personal dosemeter to give a more reliable estimate of the true dose. Circumstances in which this may be necessary include:

- when a whole body dosemeter is worn over a lead apron in order to compensate for the protection offered by the lead apron (e.g. a factor of 0.1, see section 4.6.1);
- eye dosimetry, in order to compensate for lead glasses (e.g. a factor of 0.5, see section 4.6.4.2);
- hand dosimetry (worn at different positions), in order to compensate for the large dose gradients across a hand (the correction factor is variable, see section 3.2.2.4).

Even if the values of these factors are agreed nationally so that a consistent approach is applied, the application of each factor will be specific for different wearers and may need to be varied depending on the circumstances. It is up to the employer and employer's RPA to decide on suitable correction factors for a specific wearer, not the ADS.

ADSs report measured doses and have not historically applied these corrections. An ADS cannot start applying correction factors to the assessed doses or use algorithms to combine over and under lead apron measurements, unless this has been approved by HSE. The employer must have a systems in place to confirm that the dosemeter is always worn in the designated position, any relevant PPE is correctly used, so the factors applied remain relevant for each specific wearer.

Applying corrections factors will ensure that doses for classified workers reported to the Central Index of Dose Information (CIDI) are a more realistic assessment of the true dose. ADSs' computerised systems will need to be significantly altered to enable appropriate factors to be stored for each wearer and to introduce flags to indicate when the factors have been applied.

Assessments carried out with a collar dosemeter (i.e. for non-classified staff) would not form part of the approved service undertaken by an ADS. Any dose assessments entered into the dose record kept by a coordination and record keeping service would need to be entered as estimates only.

8.5 Monitoring pregnant staff

For women who are not pregnant, the same dose restrictions apply as for men. However, for a woman who is, or may be pregnant, additional controls are necessary to restrict the dose to the unborn child. IRR17 Regulation 9(6) specifies a 1 mSv dose constraint for pregnant staff. Although not strictly a dose limit, this means that once the employee has notified the employer that she is pregnant or the employer should reasonably be expected to be aware of the pregnancy, the employer must ensure that the equivalent dose to the foetus is unlikely to exceed 1 millisievert (mSv) during the remainder of the pregnancy. IRR17 Regulation 15 requires that employees are trained so that they are aware of the need to declare pregnancy.

The employer must carry out risk assessments that consider the potential exposure of the foetus and highlight the need for specific restrictions in working practice. In most cases personal radiation monitoring will already be being undertaken, but the risk assessment may highlight a need for personal monitoring if this is not already being carried out or an adaptation of current practice. In most cases passive personal monitoring is sufficient. However, in situations where staff could receive a significant dose quite quickly due to possible prolonged high dose rates or because there is a need for more rapid knowledge of the dose (e.g. accident in PET or radionuclide therapies) the use of active electronic dosemeters might be appropriate.

Strict adherence to proper working practices incorporated into Local Rules for all staff should ensure that all doses are ALARP. Routine external monitoring of staff working within healthcare (Oatway *et al* 2016) show that less than 3% of staff have reported doses exceeding the public ionising radiation dose limit of 1 mSv per year. Certain categories of staff, particularly those involved in interventional x-ray work, cardiac catheterisation laboratories and nuclear medicine procedures, including positron emission tomography (PET) may exceed the 1 mSv constraint if they do not modify their working practices.

For exposure to external radiation, a restriction of 1 mSv dose to the foetus can be taken as broadly equivalent to a dose to the surface of the abdomen of a pregnant woman of about 2 mSv when exposed to diagnostic x-rays (HSE 2018). Faulkner and Marshall (1993) measured the ratio of the uterus dose to the whole body personal monitoring dose recorded under the lead apron for various monitor positions, different lead apron thicknesses and different x-ray tube positions. Their results, confirmed by ICRP (2000), demonstrated that the dose recorded by a dosemeter worn at the neck rather than under a lead apron will overestimate the foetal dose by at least a factor of 33. It was recommended in section 4.6.1 that when staff are routinely monitored with a single collar dosemeter worn outside the lead apron then a second dosemeter under the lead apron should be worn if the $H_p(10)$ reading exceeds 6 mSv (figure 4.2). This approach should also ensure any foetal dose remains less than 1 mSv provided the lead apron is being worn appropriately and is not damaged.

> Compliance with the dose constraint for pregnant staff working in diagnostic radiology is best demonstrated with a personal dosemeter worn at the waist but under the lead apron. A constraint of 2 mSv to the abdomen can be applied. A collar dosemeter should be sufficient for staff undertaking occasional work involving x-ray exposure.

For work with ^{99m}Tc or ^{131}I, it has been shown that a limit of 1.3 mSv to the maternal abdomen surface will restrict the foetal dose to 1 mSv (Mountford 1997). For practical purposes and as a conservative measure it is recommended that the abdominal limit is set at 1 mSv when working with ^{99m}Tc or ^{131}I. A 1 mSv limit should definitely be applied for staff working with higher energy radiation such the use of ^{18}F in PET imaging or in radiotherapy.

> A limit of 1 mSv limit to the abdomen should be applied to pregnant staff working in nuclear medicine and radiotherapy.

The exposure to the foetus from intake of radionuclides by the mother should also be considered when applying the dose constraint of 1 mSv. The foetal dose can be calculated using published conversion coefficients (ICRP 2001) if the mother's intake of activity of a specific radionuclide is known. Generally the coefficient (Sv/Bq) is highest for intakes at conception or early in pregnancy as is the case for ^{99m}Tc. Foetal uptake of iodine however increases rapidly about 11 weeks after conception when the thyroid becomes active so the coefficients are very much greater at the end of pregnancy. Restrictions to limit the dose to the member of staff will usually restrict the dose to the foetus to a greater extent because the coefficients for calculating the foetal dose are less than the coefficients for calculating the mother's effective dose. This is not the case for uptake of iodine and the baby's dose could be more than double the mother's dose (Phipps *et al* 2001).

8.6 Employees with multiple employers

Some individuals may work at multiple sites and have more than one employer. In this situation each employer cannot necessarily treat the doses received by the employee at their sites in isolation because it is the total dose that the individual is receiving overall which needs to be considered in relation to dose limits. Managing their dose record can be difficult and requires cooperation between the employers (Reg 16, IRR17) and the employee. Employers will need to know if their employees work with different employers and will need to have systems in place capable of capturing doses received where this work involves interventional radiology.

The exact arrangements may depend on the expected level of dose determined by the risk assessment. HSE have issued guidance and some of the recommendations given are summarised below (HSE 2003a). A BIR led working group of professional bodies has recently produced guidance (Rogers *et al* 2017) specifically on this subject with support from HSE.

The personal monitoring management system discussed in section 8.1 needs to also consider:

- how the cooperation between employers will work;
- whether staff wear a single set of dosemeters or a set for each employer;
- how dosemeter results will be shared if only a single set is worn;
- how dosemeters results will be collated if multiple sets are used;
- which employer investigates breaches of investigation levels, how conclusions of an investigation will be shared and who takes responsibility for required actions;
- how will employers determine at which site any dose was received.

If the doses that staff are likely to receive are low (e.g. < 10% of any dose limit), the simplest solution is for one employer to issue a dosemeter which is worn at all times. This avoids the need for employees to keep switching their dosemeters and therefore reduces the possibility of mixing them up or forgetting to wear them. Conversely, there is the risk of them leaving the dosemeter at one employment and not therefore wearing it at the next. If a significant dose is recorded on the dosemeter or a dose investigation level is breached each employer should issue their own dosemeters to establish the source of the unexpected dose. Full cooperation between all the employers would be required for the resultant investigation. If the risk assessments demonstrates that the total dose received is between 10% and 30% of any dose limit, more careful consideration of whether to monitor separately at sites is required. It is not advisable to monitor separately at sites where the dose is likely to be below relevant dose measurement thresholds. However an electronic personal dosemeter may help to confirm periodically that dose remain low at particular sites. An employer (often the main employer) may issue a dosemeter for use at all times and expect other employers to issue additional dosemeters for use on their premises only. The employee will then have a number of dosemeters and possibly dose records, but one record will give the total dose received, which can be shared between employers if necessary. If the radiation risk assessment concludes that the member of staff must be classified, a separate set of dosemeters is more likely to be required. It is worth noting that if a worker exceeds the dose limit, it is the employer at the particular point when the dose limit is exceeded that is legally responsible (for example, even if the employee only receives an eye dose of 0.2 mSv working for them and the remaining 19.9 mSv comes from work with another employer). Every employer must be aware of exposures received elsewhere to make sure dose limits are complied with. For workers who are at risk of approaching a whole body effective dose limit the use of an EPD to record the cumulative dose might provide a useful check.

Employers will need to have a system in place (e.g. the radiation risk assessment, especially for new practices) to check if any employees with multiple employments have been designated as classified persons by another employer or whether the total annual dose requires classification. For staff who are classified, each employer must make arrangements either individually or jointly, for monitoring by an ADS. The member of staff may also be a classified outside worker (see below) and so must be issued with a passbook. It is worth noting that someone with several employers can be classified and never be an outside worker provided they are always directly employed in each facility. Employers may decide to issue their own set of dosemeters for use solely on their premises and arrange for an ADS to maintain a dose record—as a minimum they must be able to assess the dose to enter into the wearer's passbook. Employees with more than one employer may have a number of dosemeters and dose records, possibly from different dosimetry services. Each employer must consider the total dose received by the employee (to demonstrate that a dose limit will not be exceeded) and to set and identify whether an investigation level or classification level has been reached. For this reason each ADS may be asked to issue dose reports to the other employers (once the wearers have granted permission for this exchange of information to take place). Such methods will work only if there is full cooperation between employers.

8.7 Outside workers

Under IRR17, an outside worker (OW) is *any* exposed worker who carries out services in the controlled or supervised areas of any employer (other than those of their own employer). The regulations include both classified and non-classified outside workers. Under IRR99, an OW was defined as *a classified person* who carries out services in the *controlled* area of any employer (other than the controlled area of his own employer). So the new definition of OW is broader including non-classified workers and work in supervised as well as controlled areas.

Under IRR17 a worker entering a controlled or supervised area of another employer is not an OW if:

- the site owner or occupier formally hands over responsibility for that area to the worker's employer for the duration of a specific project;
- they enter the area merely as a visitor (i.e. they are not providing services); or
- they have an *individual* contract of employment with the employer responsible for that area (as distinct from any contract for service between their own employer and the employer responsible for the area).

Various legal requirements apply to the employer who has designated the controlled or supervised area (IRR17, Reg19) including ensuring the OW has received appropriate training in the use of personal protective equipment and keeping doses ALARP and that the non-classified OW's doses are below classification level. However, the need to record individual monitoring data and issue radiation passbooks only applies to classified outside workers.

The radiation passbook carries identifying details of the outside worker and information about the assessed and estimated doses received by that worker.

The radiation passbook should contain the following information;

- individual serial number of the passbook;
- a statement indicating that the passbook is approved by the HSE;
- date of issue;
- details of the issuing ADS (name, telephone etc);
- details of the employer (including email address);
- details of the employee (name, date of birth, gender, national insurance number);
- date of last medical review;
- details of relevant dose limits;
- cumulative doses for the year to date;
- details of the employer where the work is undertaken;
- estimates of the doses received.

The employer must have arrangements to ensure that the passbook is kept up-to-date.

The following is taken from guidance produced by HSE (2003b) in which they define the OW's main employer as the 'direct employer' (DE) and the other employer who has effective control of a controlled area where an OW carries out services is known as the 'indirect employer' (IE).

The IE must ensure that arrangements are in place:

- To estimate the dose that the OW receives whilst working in the IE's controlled area. Dose estimates are needed in any situation that would require the dose received to be assessed under the regulations. This may include dose estimates to the extremities or eyes. For external whole body dosimetry an electronic direct reading personal dosemeter will be the easiest method for obtaining the dose estimate. For extremity and eye doses, the estimate may need to be obtained by multiplying the whole body estimate by a multiplicative factor. This factor could be derived from reliable formal past records for the same type of work or by experiment.
- To ensure that the dose estimate is entered into the OW's radiation passbook as soon as reasonably practical after the OW has completed work in the controlled area. For this reason the quality of information may be less important than the immediacy of updating the record. The dose estimate should be made and entered when the activities on the IE's site have ended, or when outside workers leave the site in order to take up work for a different employer. The IE does not have to enter dose estimates in passbooks each time outside workers physically leave the site (e.g. at the end of a day's work) provided that they are not going to another radiation employer before they return.

Methods of estimating internal doses and doses following a contamination incident are discussed in HSE's guidance (2003b).

A classified OW may be given separate sets of dosemeters at each site at which they work. These dosemeters may be issued by different ADSs (assessment). The coordination and record keeping ADS will need to ensure all these dosimetry results are collated into a single record for the worker.

8.8 Unusual dose readings

A dosemeter may record an unusually large dose value occasionally. If this dose is substantial, e.g. above a dose limit, the ADS will phone or email the result through to the employer. A common failing is the initial assumption that the dosemeter reading is wrong or it must have been exposed whilst not being worn. The ADS will have done its own checks to ensure that the assessed dose is 'real' and not a result of a faulty dosemeter. Initially the user should assume the dose is real and so needs to be explained. The action required depends on the individual dosemeter value and the cumulative dose:

(a) If a member of staff receives a dose reading which is slightly larger than usual (e.g. a whole body 0.3 mSv rather than the usual 0 mSv), this is not necessarily an incident. There may be a satisfactory explanation. The Departmental Manager should be monitoring all dose assessments returned by the dosimetry service and should therefore be aware of normal doses for different members of staff. If a particular member of staff receives an unusual dose the line manager should carry out a check to help identify possible reasons e.g.

- the person carried out a larger workload than normal,
- the dosemeter was accidentally left in an x-ray room or accidentally (or deliberately) irradiated, or
- the dosemeter was worn during a medical exposure or
- the dosemeter has passed through security scanning systems etc.

If it is decided that the dose in the person's record should be altered, any changes should be recorded locally.

If a member of staff receives a cumulative dose within the year greater than the dose investigation level (IRR17 Reg 9 (8)), specified in the local rules (HSE 2018) the employer must carry out a **formal** investigation. This is normally the responsibility of the Department Manager who may often involve the help of the RPS. This investigation is to check whether the exposures are being restricted to ALARP and it must be documented. Although the regulations only require an investigation level in terms of effective dose, employers may also choose to set investigation levels for extremity and eye doses.

(b) If the wearer's cumulative dose for the year, including the dose just assessed, exceeds a dose limit (i.e. an overexposure has occurred), Reg 26 of IRR17 requires the employer to undertake an immediate investigation to determine whether there are circumstances which show **beyond reasonable doubt** that no overexposure could have occurred. The purpose of the immediate investigation is to rule out incidents which can be readily shown not to have taken place. The burden of proof lies with the employer to demonstrate that the wearer has not received an overexposure (rather than the other way round). This level of proof is more stringent than 'on the balance of probabilities'. It is for this reason that it is very important that wearers notify lost, damaged, or accidentally exposed dosemeters **before** they are read by the dosimetry service (see section 1). A second common failing is to delay notification until a detailed investigation has been completed. If the immediate investigation cannot demonstrate that the overexposure has not occurred the employer should notify HSE as soon as practicable (i.e. after the immediate investigation which should normally be completed in within a day or two). Any instances of overexposures should be escalated with the organisation so that senior management are fully aware of the incident and the notification.

8.9 Overexposures and notifications to HSE

If an individual has definitely received an overexposure (or an immediate investigation cannot show an overexposure has not occurred) the HSE should be notified as soon as practicable. Notification should be via email, i.e. irrnot@hse.gov.uk within England, Scotland and Wales and mail@hseni.gov.uk within Northern Ireland.

The detailed investigation (which will usually follow the notification) is carried out to establish why the overexposure occurred, what dose was received and what steps are necessary to prevent a reoccurrence. The investigation should include (HSE 2018):

- work routine of the wearer and relevant work colleagues during the period concerned;
- personal monitoring results and results of any radiation monitors in use at the time;
- involvement with previous incidents;
- dose history of wearer and colleagues undertaking similar work;
- radiation surveys in the area concerned;
- whether local rules were being followed;
- consideration of all possible scenarios for suspected overexposure;
- training received by wearer and competence for the work undertaken;
- the need for a reconstruction of the incident to measure the dose that may have been received.

8.10 Special entries and changes to the dose record

The dose entered into an employee's dose record may be wrong for several reasons including:

- the employee was exposed when they were not wearing a dosemeter;
- the dosemeter reading is not a true representation of the actual dose received (e.g. ring monitor in CT guided interventions);
- the employee reveals a dosemeter was exposed to high dose rates when not being worn or they were wearing the wrong dosemeter.

For classified wearers, the formal dose record is a legal requirement and an amendment to the dose record because the dose is much greater than or much less than recorded counts as a 'special entry'. Special entries are appropriate (HSE 2018) if the dose received differs from the original entry in the dose record by:

- at least 1 mSv (for recorded doses $\leqslant$ 1 mSv) or
- a factor of $\geqslant$ 2 (for: 1 mSv < recorded doses < the relevant dose limit) or
- a factor of $\geqslant$1.5 (for recorded doses $\geqslant$ the relevant dose limit).

The employer can make arrangements for their ADS (records) to make a special entry when the above criteria are met provided they have done an adequate investigation. HSE have provided detailed advice on special entries and what they would consider as an adequate investigation (HSE 2000). The ADS will require a summary of the investigation that was carried out to estimate the actual dose received which is to be entered into the dose record. Special entries can be made without HSE approval except when the cumulative dose in the wearer's dose record is greater than any dose limit (IRR17 Reg. 23(8)).

As there is no legal obligation to maintain dose records for non-classified wearers, values in any dose records that are being kept can be changed as appropriate. The ADS should be advised if the dose assessment was not valid and a better estimate of the dose should be provided to go into the member of staff's dose record. However, it is prudent to involve HSE when proposing to remove doses which are more than the dose limit.

References

Faulkner K and Marshall N W 1993 Personal monitoring of pregnant staff in diagnostic radiology *J. Radiol. Prot.* **13** 259–65

HSE 2000 The Ionising radiations Regulations 1999: Special entries in dose records. *HSE 560/43* http://www.hse.gov.uk/radiation/ionising/doses/560-43.pdf (accessed 21/3/16)

HSE 2001 Radiation Protection News Issue 20, Personal dosimetry: the duties of the employees http://www.hse.gov.uk/radiation/rpnews/rpa20.htm#personal-dosimetry (accessed 21/3/16)

HSE 2003a Radiation Protection News Issue 23. Dosimetry for employees with more than one employer http://www.hse.gov.uk/radiation/rpnews/rpa23.htm#DOSIMETRY (accessed 21/3/16)

HSE 2003b *The Ionising Radiations Regulations 1999. Outside Workers (Including applications for exemption from issue and use of radiation passbooks)* http://www.hse.gov.uk/foi/internalops/ocs/500-599/560_54.htm (accessed 21/3/16)

HSE 2018 Work with Ionising Radiation. Ionising Radiations Regulations 2017, Approved Code of Practice and Guidance Publication L121 (second edition) (London, UK: Health and Safety Executive, The Stationery Office). http://www.hse.gov.uk/pubns/books/l121.htm (accessed 18/9/2018)

ICRP 2000 Pregnancy and medical radiation. ICRP Publication 84 *Ann. ICRP* **30** 33–4

ICRP 2001 Doses to the embryo and foetus from intakes of radionuclides by the mother. ICRP Publication 88 *Ann. ICRP* **31**

Mountford P J 1997 Risk assessment of the nuclear medicine patient *Br. J. Radiol.* **70** 671–84

Oatway W B, Jones A L, Holmes S, Watson S and Cabianca T 2016 Ionising radiation exposure of the UK Population: 2010 Review *Publication PHE-CRCE-026* Public Health England, Chilton, UK https://www.phe-protectionservices.org.uk/cms/assets/gfx/content/resource_3595csc0e8517b1f.pdf (accessed 12/8/16)

Phipps A W, Smith T J, Fell T P and Harrison J D 2001 Doses to the embryo/foetus and neonate from intakes of radionuclides by the mother. Part 1: Doses received in utero and from activity present at birth *CRR 397/2001* Health & Safety Executive, London, UK http://www.hse.gov.uk/research/crr_htm/2001/crr01397.htm (accessed 27/2/2018)

Rogers A, Chapple C L, Murray M, Platton D and Saunderson J 2017 UK guidance on the management of personal dosimetry systems for healthcare staff working at multiple organizations *Br. J. Radiol.* **90** 20170363

www.ingramcontent.com/pod-product-compliance
Ingram Content Group UK Ltd.
Pitfield, Milton Keynes, MK11 3LW, UK
UKHW051849100926
8556IPUK00005B/536

9 780750 321976